Invasions of the Land

Invasions of the Land

The Transitions of Organisms from Aquatic to Terrestrial Life

Malcolm S. Gordon and Everett C. Olson

Columbia University Press

mah *New York*

Columbia University Press
New York   Chichester, West Sussex

Copyright © 1995 Columbia University Press
All rights reserved

Library of Congress Cataloging-in-Publication Data

Gordon, Malcolm S.
   Invasions of the land : the transitions of organisms from aquatic
to terrestrial life / Malcolm S. Gordon and Everett C. Olson.
     p. cm.
   Includes bibliographical references (p.   ) and index.
   ISBN 0–231–06876–X
   1. Evolution (Biology) 2. Adaptation (Biology) I. Olson,
Everett Claire, 1910– . II. Title.
QH366.2.G657 1994
575—dc20   94–20344 CIP

Printed in the United States of America

c 10 9 8 7 6 5 4 3 2 1

With love and gratitude for inspiration, support, partnership—
and much more, both in the past:
Marjorie J. Smolensky Weinzweig (1935–1990)
and in the present:
Carol Ann Gordon

Malcolm S. Gordon

In Memoriam
Everett C. Olson
1910–1993

Esteemed coauthor, colleague, mentor, and friend, Ole passed away while this book was in press. He was a distinguished paleontologist, geologist, and creative student of evolutionary biology. His contributions to this book represent his mature overview of a major area of paleontology that was a focus of his interest throughout his career. We miss him.

Malcolm S. Gordon
David J. Chapman

Contents

Preface

This book has two primary goals:

- To provide an up-to-date, comprehensive overview of what is known and not known about the evolutionary transitions made by plants and animals as they moved from aquatic to terrestrial environments.
- To place this factual picture in a conceptual and theoretical framework that is significantly different from those used in existing descriptions of the subject.

This book differs from previous treatments in four specific ways. First, it is multiply authored yet internally integrated. Our colleague David J. Chapman has graciously contributed a chapter on invasions of the land by plants. Each of the three authors has worked for many years on research problems relating to the subject, but in different and complementary ways. Gordon is an evolutionarily oriented animal physiologist specializing in ecological physiology. Olson is a geologist and paleontologist; Chapman is an algalogist and a student of plant evolution.

Second, it gives equal emphasis to paleobiology and neobiology. Third, it incorporates the newest information relevant to the water-land transitions deriving from recent applications of micropaleontological and molecular methods and applies this new information to classical issues in the field.

Fourth, and finally, it evaluates the evidence relating to each transition without reference to specific hypothetical scenarios. Removing prior constraints on thinking leads us to several new perceptions—notably that evidence now available is inadequate to permit the description of the actual historical transition for *any* of the living terrestrial groups (plant or animal); that the vertebrates are the only group for which even the outlines of plausible scenarios can be constructed; and that, for the vertebrates, a wide variety of scenarios are plausible, including multiple origins of tetrapods.

The subject matter covered is so diverse, complex, and voluminous that we have made no effort to be encyclopedic in terms of the literature. We have striven instead to produce a readable and coherent summary account of current understanding of the main events and processes involved in the transitions from water to land of all the major groups of plants and metazoan animals for which at least some direct evidence exists. To this end, many references pertain to review and summary publications. Much of the text, however, is based on cited papers in the primary research literature.

A list of section heads appears at the beginning of each chapter. The overall format is that of a series of selective critical reviews directed primarily at audiences of professional colleagues and advanced students. Nevertheless, we hope our work will also be accessible to scientists working in fields other than paleobiology, physiology, systematics, and evolution, who may profit from a deeper understanding of these critical episodes in the history of life.

This book arose out of the fortuitous happenstance that the three of us became colleagues in the Department of Biology at the University of California (Los Angeles) twenty-five years ago, where we have remained ever since. Fifteen years later the idea for this book was born. The book thus shares with its subject matter such features as a long, complex history, multiple origins, and the workings of a good deal of chance.

Each chapter notes the particular author(s) whose expertise and views were primarily expressed in that part. Nevertheless, shared philosophical and theoretical frameworks and approaches evolved among us in the course of our collaboration. These are articulated specifically in chapters 1 and 10 and are manifested throughout the text. Gordon acted as overall editor, coordinator, and gadfly.

Many people have assisted us in the long course of this project. We are grateful to all. We wish to acknowledge especially C. Barnhart, M. Feder, C. Gans, J. Graham, G. Lauder, L. Mantel, B. McMahon, R. McMahon, D. Wolcott, and T. Wolcott. Several anonymous reviewers were also very helpful. The original art work was drawn by K. Bolles and M. Kowalczyk. Major parts of the manuscript were typed by B. Phan, T. Zeyen, and K. Brill. E. Lugenbeel and his staff at Columbia University Press were cooperative and supportive. Financial assistance for various parts of the project derived from research grants from the Biology-Fisheries Program and the Academic Senate Committee on Research at the University of California, Los Angeles.

Invasions of the Land

1
Introduction

MALCOLM S. GORDON, EVERETT C. OLSON,
AND DAVID J. CHAPMAN

1.1 Scope and Organization of Chapters
1.2 Definitions and Limits
1.3 Previous Studies
1.4 Concepts and Key Assumptions

Few developments in the evolutionary history of life are as important to primarily terrestrial mammals like us than the transitions made by some groups of plants and animals from aquatic to terrestrial modes of living. This book is about these transitions, specifically:

1. What is known and not known factually about the transitions, including relevant evidence from both the fossil record and living organisms; and
2. New or different theoretical and conceptual insights relating to evolutionary processes that can be derived from the study of these transitions.

On the factual level this book considers the questions: When did the transitions occur? Where did they occur? Which groups of organisms were involved? What were the environments in which the events took place, and what were the physical conditions there and then? What were the biochemical, physiological, morphological, behavioral, and ecological properties of the lineages in transit that permitted this radical evolutionary step? How did these lineages change in the course of the transition and in its aftermath? What major interactions occurred between organisms and their environments and between major groups of organisms venturing onto land?

Many approaches have been taken over the years in answering these questions—at least in providing partial answers. We feel that enough new information has recently come to light that another comprehensive overview is in order, to integrate and synthesize the ideas and concepts in the diverse and scattered literature and to point out remaining gaps in knowledge and understanding. Our aim is to summarize and evaluate what is known and not known. But we also believe that the topic is ripe for an alternative theoretical framework through which to view and interpret the facts.

Notably, viewing the facts on their own, independent of many traditional assumptions, led us to a nontraditional approach to the interpretation of important features of the fossil records and evolutionary histories of major groups. Our broad survey of many of the functional properties and adaptations for amphibious lifestyles possessed by living organisms has led us to a revisionist view of the credible possibilities for the paleoenvironments and the paleoclimates within which the historic transitions might have occurred. Finally, by modifying and changing the emphases in some important assumptions, we believe that our study of the transitions opens up possibilities for similarly new perspectives on other major evolutionary developments.

1.1 Scope and Organization of Chapters

The factual parts of this book present a series of selective and critical reviews that have been integrated and coordinated to form a coherent whole. The perspectives taken in these reviews are intended to present a consensus of current opinions among active workers in the various disciplines, modified by the personal judgments of the present authors. Alternative opinions, however, are presented where relevant.

The overview starts with geology, paleoclimatology, and paleoecology (chapter 2). It then proceeds to paleontology, describing the fossil records for both plants and animals (chapters 3–6). The circumstantial evidence from living plants and animals is also summarized (chapters 3, 7–9). The book ends (chapter 10) with a summary of the most important findings from the factual overview. This is followed by an attempt at synthesis and new interpretations. The synthesis presents a theoretical and conceptual framework that yields:

1. A fresh way of looking at the fossil records of the transitions and of interpreting the significance of what is found in those records;
2. New sets of alternative scenarios that indicate how the transitions might have occurred in the two major groups for which we have substantial factual information, the plants and the vertebrates;
3. Our suggestions for possible additional applications of the approaches developed here.

1.2 Definitions and Limits

The story of the transitions of multicellular forms of life to terrestrial modes of living need delve no further into the past than the late Precambrian. Our discussion rests upon the assumption that the organisms directly ancestral to those that occupied the land developed in aquatic environments. Thus the evolutionary transitions from water to land were primary in the lineages so involved.

With respect to plants, we focus on vascular plants. We say little concerning the evolution of bryophytes, as bryophytes are not ancestral to vascular plants. Our use of the word *plant* includes their algal forebears. Our occasional use of the word *animal* refers strictly to the metazoans.

We define *aquatic organisms* as those that carry out all aspects of their life histories, at all stages of their development and throughout their geographic ranges, entirely within water of some kind. *Fully terrestrial organisms* similarly live only in subaerial environments. These definitions clearly will not apply to all members of many groups of organisms, groups here referring to systematic categories (taxa) from the levels of genera to phyla. The definitions specifically exclude the quasi-terrestrial organisms that live on or in the land and that form the broad category here termed *cryptics* (also called cryptozoic animals by some authors). For at least part of their lives cryptic organisms, such as many nematodes and oligochaete earthworms, exist in an aquatic medium, mostly within sediments along marine or estuarine shorelines, in rocky coastal seaweed mats, or in soils or in forest floor litter (see chapter 4 for fuller discussion of these definitions). Recent reviews relating to cryptic organisms were written by Chapman (1977), Higgins and Thiel (1988), and Brown and McLachlan (1990).

The taxa that we consider to be *terrestrial* are those in which substantial numbers of species are fully terrestrial. A comparable definition applies to *aquatic* taxa. The lower and upper limits of our consideration of both organisms and taxa are reached when they are, respectively, fully aquatic or fully terrestrial. *Amphibious* taxa live, to varying extents, both aquatically and terrestrially. We focus primarily on the early stages of the various transitions, as terrestrial radiations are well reviewed in many other books and articles.

The second major factor that determined our scope of study is the current state of knowledge of the organisms. We decided to emphasize those organisms and groups that (1) have been most successful in invading the land and becoming fully terrestrial, (2) have been best studied in that context, and (3) have some significant evidence relating to their transitions. For the plants these groups are the green algae and the psilophytes. For the animals they are the gastropod molluscs, the gnathobdellid leeches, the several groups of crustacean arthropods, and the vertebrates. We omit from full consideration the insects, for which no relevant evidence of the transition now exists, and the variety of groups (such as water beetles and cetaceans) in which secondary reinvasions of aquatic environments have occurred.

These two policy positions, the definitions and the limits, justify only cursory treatment of the cryptics. Some cryptics, notably nematodes and various crustacean groups, may well have played important roles in the earliest stages of the occupation of the land by multicellular organisms. Many of them arguably were, or are, to some extent aquatic. Only recently have bits of data from the fossil record—primarily from fossil soils—begun to give evidence of the roles of cryptics in early occupancy of land (chapters 3 and 4). Several successful terrestrial taxa, notably terrestrial gastropods, leeches, crustaceans, and the insects, have no known aquatic ancestors or transitional primary amphibious forms in the fossil record. These taxa may well have made the transition via the cryptic fauna. Our judgment, rather, is that this is one of several large subject areas needing further study.

We consider only those properties of organisms that relate unequivocally to the aquatic-terrestrial transition as it occurred in the organisms' own taxon. A consequence of this is that our emphasis falls primarily upon the whole-organism level of structural complexity. However, where appropriate we include information from all relevant levels of complexity, from the molecular to the ecosystem.

At the organismic level the multiple differences between aquatic and terrestrial environments have required multiple and complex changes involving almost all major organ systems. The principal considerations with respect to *plants* are discussed in detail in chapter 3. The list includes:

- *Desiccating environments on land,* requiring waterproofing.
- *Changed requirements for aerial gas exchanges,* including water vapor exchanges, largely produced by the necessity of waterproofing.
- For at least the aerial parts of land plants, *deficiencies of water and nutrients in terrestrial habitats.*
- *Changed requirements for mechanical support.*
- *Increased needs for transport mechanisms* for water, nutrients, and photosynthetic products between roots and the aerial parts of plants.
- *Changed requirements for successful reproduction.*

The major differences we consider in the context of each *animal* taxon are as follows (see also Dejours 1987).

- *Mechanical support and the effects of gravity.* Aquatic organisms rely upon dense and buoyant media; terrestrial organisms lack that support.
- *Water and land as environments for locomotion, activity, and dispersal.* Both passive and active movements are affected by the differences between hydrodynamics and aerodynamics, between drag and friction.
- *Water and air as respiratory media.* Differences in density and viscosity, in the concentrations of the major respiratory gases, and in the diffusion rates of these gases in the two media have produced a range of major differences. Mechanisms of acid-base regulation have also been significantly affected. Changes in respiratory mechanisms have often involved changes in circulatory systems as well.
- *Water and air as media for thermal exchanges.* Differences in heat capacities, conductivities, and convective properties have led to important differences in organisms.
- *Abundance and availability of water in the two environments.* The processes involved in maintenance of adequate levels of body hydration are very different.

- *Abundance and availability of necessary inorganic solutes in the two environments.* Maintenance of proper internal concentrations of important solutes involves different processes and organs.
- *Biochemical constraints on the production and disposal of metabolic wastes, especially nitrogenous wastes, in the two environments.* There has clearly been substantial coevolution of mechanisms of waste handling and disposal in relation to both water and solutes.
- *Water and air as information transporting media.* The major sensory modalities of animals operate differently in the two environments. Orientation mechanisms to gravity, to geographic locale, to potential mates or predators are all different.
- *Water and air as media for reproduction.* Physical and chemical differences between water and air have fostered a myriad of differences in water- or air-adapted organisms at the biochemical, physiological, and behavioral levels.
- *Water and air as differential selectors for life history patterns and modalities.* Behavioral and ecological relationships have also often changed.

1.3 Previous Studies

We consider ourselves fortunate that, during the years when this book was only an idea in our minds, Colin Little was working actively on his major reviews of many parts of the relevant literature. We see no value in trying to duplicate his excellent and encyclopedic works, *The Colonisation of the Land* (1983) and *The Terrestrial Invasion* (1990). Rather, this book looks at the facts from several different perspectives, and in doing so it covers in detail a range of subjects that Little addresses only briefly or not at all. The largest and most important of these are the geological background, paleoclimates, and the fossil records and paleoecology of animals and plants. We also focus on living quasi-terrestrial plants. For the primarily zoological topics that Little does cover we emphasize different theoretical approaches, other aspects of the literature, and more recent developments—including many deriving from molecular biology. The attention we give to evolutionary theory and process also distinguishes our effort from that of Little.

Considerable material relevant to our discussions is included in two recent books (Burggren and McMahon 1988; Feder and Burggren 1992) and in the proceedings of a major symposium devoted to comparing and contrasting the physiological properties of aquatic and terrestrial animals (Dejours et al. 1987). Additional recent reviews are by Selden and Edwards (1989), Labandeira and Beall (1990), Shear (1991), and Gray and Shear (1992).

1.4 Concepts and Key Assumptions

Our overall approach is based on four primary concepts and assumptions. Since some of the technical terms we use have multiple possible meanings (Keller and Lloyd 1992), we try here to make the component definitions explicit.

1. The principle of *uniformitarianism*, both generally and specifically, applies to the interpretation of the fossil records of the transitions. General applicability (i.e., the laws of physics and chemistry are invariant over time) requires no further comment—the history of life cannot otherwise be scientifically understood. Specific applicability, in the context of this book, means:

(a) We believe that the physical-chemical *complexity* of paleoenvironments since at least the Cambrian was qualitatively indistinguishable from the physical-chemical complexity of contemporary environments. Thus, for example, Silurian or Devonian rocky coastlines, beaches, estuaries, ponds, lakes, streams, and rivers provided diversities of both microhabitats and macrohabitats comparable to those found today in equivalent environments. These paleoenvironments were also comparable in variability, both temporally and spatially, to their present-day counterparts.

(b) The *physical sizes and extents* of the marine and freshwater shorelines of the paleocontinents were comparable to present-day shorelines. In other words, thousands of kilometers and hectares of habitat suitable for transitions existed.

(c) *Functional uniformitarianism* applies to all organisms. The biochemical, physiological, morphological, and behavioral diversities, plasticities, and adaptabilities of the populations of paleoorganisms living in the inshore marine, littoral, and inland waters of the paleocontinents were also comparable to those of modern organisms.

(d) *Functional uniformitarianism is, however, the only subtype of uniformitarianism relating to organisms that we espouse.* We disavow any implications or inferences relating to possible broader interpretations, such as taxonomic or ecological uniformitarianism over long spans of time. Notably, Paleozoic paleoecology appears to be a specialized field of its own. The ecologies of Paleozoic habitats, such as salt or freshwater marshes, were clearly different from the ecologies of those habitats today. Knowledge of present-day ecology is of limited value for developing an understanding of the workings of Paleozoic habitats and ecosystems.

2. *We believe that the probability is either zero or very small that, for each major group, there was only a single transition—that only a single subgroup (or population) emerged from water onto land and in a single location.* We believe instead that over extended periods of time—in various locations and in different regions widely scattered over the surfaces and around the edges of the paleocontinents—many different subgroups made transitions. These adventurers probably belonged to several or perhaps many different species, genera, or even families of ancestral organisms. The selective pressures, both abiotic and biotic, that drove the transitions varied widely at different times and in different places, as did the subgroups on which they operated. Genetic interchange between geographically distant populations probably occurred to varying degrees in different lineages; in many cases such interchanges may have been zero. The effects of selective pressures on organisms were, of course, constrained by the basic natures of the organisms. The overall results were large and multiple sets of transitions, with evolved taxa generally resembling one another (parallelisms due to preexisting ontogenetic, morphogenetic, and physiological similarities, plus convergences) even though widely geographically dispersed and strongly polyphyletic. Our assumption of polyphyletic origins of some of the upper levels of current classification would, from a cladistic standpoint, mean that further divisions of these taxa would have to be made.

3. The lack of even circumstantially possible fossil evidence of ancestral aquatic forms for many modern amphibious to terrestrial groups (e.g., amphipods, isopods, decapods, hexapod uniramians) makes it probable that *there were ancestral stocks that remain undiscovered or that vanished without a record.* Such groups might have neither obvious fossil derivatives nor identifiable living descendants. The continuing series of new, putatively gap-filling discoveries of fossils

that frequently derive from the application of modern paleontological methods seem strong support for this idea.

4. *It is likely that, at least in the context of water-land transitions, the fossil records of various groups may be complex conglomerates of segments of multiple sequences rather than single sequences.* We believe it is probable that each major group that was ultimately successful had separate lineages evolving in parallel that became amphibious—or even terrestrial—but then failed. We have no reliable way of identifying such partial or temporary ventures. Chances seem good that significant fractions of the known fossils that appear to relate to the successful transitions are actually remains from what might be termed false starts. Some of the extinct groups of early plants, land snails, and amphibians now considered components of evolutionary main lines may not belong there.

In chapter 10 we combine these four concepts and assumptions with observations deriving from our factual surveys (especially relating to living organisms) presented in earlier chapters. Overall, we attempt to produce a theoretical and conceptual framework that is free from monophyletic and other assumptions in order to forge an alternative (and we believe equally if not more plausible) understanding of how the various transitions might have occurred. We also present there what we consider to be some of the more important implications of this framework.

The factual surveys and analyses in chapters 2–9 are also underlain by two more limited assumptions and inferences: (1) most models of paleoclimates and paleoenvironments have wide margins for interpretational error; (2) evolutionary processes during aquatic-terrestrial transitions were primarily gradualistic. We see no need to postulate catastrophic or saltational models, but concede that some mixed models might apply to specific situations.

Our outlook throughout the book is adaptational, but in a broadly holistic rather than a "trait-function" framework. The processes of change surely were affected by the constraints and opportunities afforded by the preexisting biochemical, physiological, structural, behavioral, and ecological adaptations of aquatic organisms that made possible the initiations of transitions to subaerial existence. Such features may often have served as preadaptations or exaptations, in the sense of Gould and Vrba (1982). Preadaptations may have involved the total organism or only parts. We recognize that for each lineage at all stages before, during, or after a transition there existed an emerging suite of potential niches, of which only a small portion was realized during the coevolution of organisms and environments. Finally, attainment of full terrestriality opened a vast array of potential niches to each lineage, some small portion of which was realized in the radiations that followed successful transitions. At least for plants these outcomes were strongly influenced by coevolutionary processes involving both fungi and insects.

For each organismic group considered there are potentially three major lines of evidence: the fossil record (to the extent there is one); living organisms considered to be phylogenetically related to fossil forms that are thought to have been involved in the actual transitions; and living organisms currently occupying habitats (and having life histories) that might reasonably be considered similar to one or more of the habitats and ways in which the actual transitions might have occurred. The features important in this latter context may or may not have phylogenetic significance—that is, most living primarily amphibious forms are not demonstrably descended from the ancestral transitional organisms in their taxa.

An important consequence of the necessity of making use of large amounts of circumstantial evidence is, we believe, that it is currently impossible to make firmly based, unequivocal statements concerning what actually happened, and how it happened, in any of the transitions. We do not, therefore, regard our interpretation as beyond doubt. But we hope that readers will come to agree with us that it is at least as (if not more) plausible than the existing scenarios drawn from other—notably monophyletic—assumptions.

Information derived from living species provides an *envelope of possibilities* that may or may not include what actually occurred. Since fossils provide only limited opportunities for the study of biochemistry, physiology, behavior, and ecology, we must in those areas make do with what we can get. We must turn to the present. But we must bear in mind that the living amphibious organisms in most taxa that have successfully invaded the land are probably at evolutionary dead ends. This is likely to be true for both primarily and secondarily amphibious forms. For each group few, if any, potential terrestrial or aquatic niches exist at present.

These considerations also form the basis for one of the primary ways in which this book differs from both of the recent books (1983 and 1990) by Colin Little. Little structures most of his discussions around what we consider to be informed speculations concerning possible ancestral taxa for land animals and the paleoroutes these taxa may have followed in invading the land. He bases these speculations on particular interpretations of data relating to selected putatively representative living forms. The data are primarily ecophysiological but also include information about geographic distributions, behaviors, and ecologies.

We are substantially more conservative in these matters. As noted, we are doubtful of the phylogenetic significance of any particular patterns of amphibious or semiterrestrial adaptation by living forms. We know of no convincing evidence that the (usually) small numbers of species of living amphibious forms that have been studied, within specific taxa, are a complete representation of the diversity of adaptive patterns that exist within these taxa. Even assuming they might be truly representative, there still is no sound basis for selecting particular patterns as more informative than other patterns in the context of past transitions to the land.

The temptation to produce plausible accounts is strong. However, our perception is that the present state of knowledge permits at best just a series of possible alternative hypotheses. We describe our own hypotheses in chapters 3 and 10.

Because of our polyphyletic and other assumptions, our overall approach to the subject of the evolutionary transition to land does not lend itself to cladistic treatment in either form or terminology. This noncladistic approach also owes to three considerations relating specifically to water-land transitions:

First, in our judgment the databases available for each of the groups we discuss are inadequate to support meaningful cladistic analyses.

Second, present evidence from the fossil record within the various groups demonstrates either (a) that living amphibious and terrestrial forms are sufficiently different from those fossil forms considered closest to the historical transitions that almost no plausible evolutionary relationships can be established, or (b) that living forms are morphologically indistinguishable from aquatic relatives, so there is no way to identify specific possible antecedents. In either circumstance, cladistic methods are not helpful.

Finally, our concerns relate primarily to the multiple *functional* aspects of the transitions, considering the similarities and differences between organisms in water and on land in each major group. We are not concerned with trying to establish specific phylogenetic sequences or trying to map the distributions of particular characteristics on such sequences. Thus we consider the major goals of cladistic analyses not to be germane to our concerns.

2

The Physical Settings of Land Invasions

EVERETT C. OLSON

2.1 Timing of Land Invasions

2.2 Composition of the Atmosphere
    2.2.1 An Ozone Screen for Ultraviolet Radiation
    2.2.2 Carbon Dioxide and the Greenhouse Effect

2.3 Geography, Topography, Sediments, and Climates

2.4 Changes in Day Length

2.5 Survey of General Conditions: Proterozoic to Permian

2.6 Summary

The history of life on earth covers a time span of some 3.0 to 3.5 billion (10^9) years, but only during the final one-seventh of this time are traces of terrestrial life preserved in the fossil record as known today. The questions posed by this timing of events are many and have led to extensive investigations of the constraints posed both by the physical and biological environments and by the nature of the organisms. Interactions between evolving metabolic pathways and the physical evolution of the earth's crust eventually produced circumstances under which occupation of the continents became possible. Of primary interest is the question of the timing of the initial land invasions. The biology of extant organisms provides critical information, but data from the fragmentary records of fossil organisms—dating back to about 1 billion years ago (bya)—are equally crucial in the search for answers. Conclusions that involve interpretations of the congruences of these sources of information are inevitably somewhat uncertain and must be based on assumptions that appear to be sound but are not subject to precise testing.

One such assumption made throughout this book (and almost universally accepted) is that all terrestrial multicelled organisms ultimately arose from aquatic stocks. Supporting this assumption is the fact that all known fossils of organisms that lived earlier than about 400 million years ago (mya) have come from sediments formed in ocean waters. Furthermore, all extant terrestrial multicellular organisms can be traced back through their phylogenetic pathways to aquatic predecessors. On the other hand, some hypotheses place the ultimate origin of life on land rather than in oceanic waters. Such hypotheses, which have found limited favor among students of the origin of life, do not imply that terrestrial lineages of animals and plants developed directly on land following terrestrial origin of living matter. Rather, aquatic organisms developed after the origin. Such debates are currently unresolvable and only tangentially germane to the discussions here.

2.1 Timing of Land Invasions

Crucial, and somewhat more tractable, are questions of just when the initial invasion of the continents took place and the corollary of why it occurred when it did (table 2.1 shows the geological time scale for the Paleozoic era). There are two aspects of this problem—one concerning the various kinds of unicellular microorganisms and the other, the multicellular macroorganisms. Currently few data on fossil terrestrial microorganisms are available, these coming primarily from inferences based on ancient soils. Yet it is essentially axiomatic that the resources necessary to support macroorganisms on land were initially formed as products from microorganisms that were able to exist and reproduce in free water in the interstices of continental sediments. The first of such organisms, we may speculate, were anaerobic.

Soils formed during the Ordovician period (figure 2.1) carry traces of burrows, presumably made by large metazoans. Retallack (1985; Retallack and Feakes 1987) speculated that the burrows could have been made by millipedes, whose earliest known occurrence is in marine rocks of early Silurian age. Spores of land plants occur in rocks of late Ordovician age. It thus appears likely that a fairly complex terrestrial ecosystem may have existed at least as early as the late Ordovician. The earliest unequivocal terrestrial macrofossils, however, are vascular plants, best known from Ludlovian (Silurian)

TABLE 2.1
Geologic Time Scale for the Paleozoic Era

Period	Epoch	Age	Date (mya)
PERMIAN	Late (Zechstein)	Changxingian Longtanian Capitanian Wordian Ufimian	245
	Early(Rotliegendes)	Kungurian Artinskian Sakmarian Asselian	
			290
LATE CARBONIFEROUS (Pennsylvanian in North America)	Gzelian Kasimovian Moscovian Bashkirian	*Very complex zoning based on invertebrate fossils*	
			323
EARLY CARBONIFEROUS (Mississippian in North America)	Serpukhovian Visean Tournaisian	*Very complex zoning based on invertebrate fossils*	
			363
DEVONIAN	Late	Famennian Frasnian	
	Middle	Givetian Eifelian	
	Early	Emsian Pragian Lochkovian	
			409
SILURIAN	Pridoli		
	Ludlow	Ludfordian Gorstian	
	Wenlock	Homerian Sheinwoodian	
	Llandovery	Telychian Aeronian Rhuddanian	
			439
ORDOVICIAN	Ashgill Caradoc Llandeilo Llanvirn Arenig Tremadoc	*Complex biostratigraphic zoning*	
			510

TABLE 2.1 (*Continued*)

Period	Epoch	Age	Date (mya)
CAMBRIAN	Late (Merioneth)	Dolgellian Maentwrogian	
	Middle (St. David's)	Menevian Solvan	
	Early (Caerfai)	Lenian Atdabanian Tommotian	
			570
EDIACARAN (VENDIAN)			
			610

SOURCE for terminology and dates: Harland et al. 1990.

sediments of Wales and from slightly older sediments in Libya. In the final epoch of the Silurian, the Pridoli, remains of similar plants have been found in the state of New York, Bohemia, Russia, and southern Britain. This relatively widespread distribution raises the question of whether there were multiple origins of land floras in several areas or a single origin in one area followed by fairly rapid dispersal.

The Silurian remains and their immediate successors, which comprise both vascular plants and invertebrate animals, must have been preceded by earlier complex systems. Even during the late Silurian, in the basal strata of the Pridoli epoch of Shropshire, England (Jeram et al. 1990), and the early Devonian, in the Emsian Rhynie Chert, diverse terrestrial animal-plant communities were present. Thereafter extensive and increasingly complex communities occur sporadically and with increasing frequency in the fossil record. The lack of clear evidence of vascular plants prior to the mid-Ludlovian occurrences is partly the result of the difficulty in interpreting isolated parts of plants as vascular or nonvascular. For both plants and animals the issue is complicated by natural processes that differentially preserve parts of biotic communities, by limited exposures of fossiliferous beds, and by accidents of sampling by paleontologists. Older sites may one day be discovered. An important point, however, is that there do exist earlier sediments in which terrestrial organisms might be expected but which have failed to produce any evidence of them.

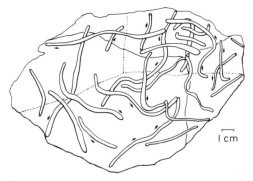

FIGURE 2.1 Ordovician trace fossils (burrows) in paleosol. Arrows indicate direction of burrowing. *After Retallack and Feakes 1987.*

Continental land masses probably date back in earth history to at least 3.0 to 3.5 bya. Land thus has been available for occupation from about the time that the first existence of organisms is documented by stromatolites. The pervasive lack of evidence of terrestrial organisms throughout most of this time, in spite of the strictures of the general incompleteness of the fossil record, must be the result of other physical and biological constraints on the invasion of land.

2.2 Composition of the Atmosphere

It has long been recognized that the origin of life and the development of various metabolic pathways in organisms were constrained and directed by the particular gases and dissolved solids in the waters of the oceans and that transitions to land were conditioned by the composition of the atmosphere. Although environment was not the only deciding factor determining the land migration of multicellular organisms, it almost certainly played a major role. It is safe to assume that the lithosphere-atmosphere system during the significant period between 650 and 450 mya was far different from that of today and was in a state of flux. The major parameters that were or could have been of significance are: (1) the partial pressure of oxygen (pO_2) and its effects on the ozone shield and incoming ultraviolet (UV) radiation; and (2) the partial pressure of CO_2 (pCO_2) and its effects on the rate of loss of heat from the earth's surface, the "greenhouse effect." Although solar luminosity in the Precambrian is believed to have been less than it is today (Levine et al. 1980), neither direct evidence nor models suggest a major departure from today's levels during the time span of the late Precambrian to late Ordovician (Kasting 1993).

Recognition that such variables as UV flux and changes in pO_2, pCO_2, and temperature could be important has stimulated efforts to place quantitative limits or boundaries on these variables and to consider the effects each might have had. This is difficult to do independent of a circular argument involving organisms. Consider: The floral and faunal changes during this time period, and particularly at the Cambrian-Precambrian boundary, suggest important environmental changes. Such floral/faunal changes include the appearance of calcification, hard parts, structural rigidity, multicellularity, and increased absolute size and surface-volume ratios. Use of these biochemical and form-function changes to argue for significant environmental alterations makes it easy, if not tempting, then to argue circularly that these environmental changes controlled or influenced the land migration. To avoid circularity in constructing the history of key environmental factors such as oxygen and carbon-dioxide, one must explore independent lines of evidence—notably, the geochemical record and modeling of atmospheric compositions.

2.2.1 An Ozone Screen for Ultraviolet Radiation

Because the incoming UV flux is regulated principally by ozone (O_3), which in turn derives from the oxygen (O_2) level, these three atmospheric variables must be considered in concert. Nucleotides absorb and may be disrupted by UV radiation in the sub-280 nanometer (nm) range. Wavelengths shorter than 320 nm are absorbed by ozone; those shorter than 280 nm are absorbed additionally by oxygen.

The importance of the ozone screen is readily apparent. Cicerone (1987) has pointed out that a 10 percent decrease in atmospheric ozone, with a full overhead sun, produces a 250 percent increase in UV penetration at 290 nm and a 500 percent increase at 287 nm. An early model by Levine (1980) illustrated the very significant dependence of the screening of incoming UV radiation on the partial pressure of oxygen in the atmosphere. Table 2.2 presents the UV screening capacity of earth's atmosphere at oxygen concentrations ranging from today's pO_2 to only 10^{-3} of it. Kasting et al. (1992) and Kasting

TABLE 2.2
*Atmospheric Oxygen and Ultraviolet Radiation Fluxes
Reaching Earth's Surface*

Wavelengths (nm)	Atmospheric Oxygen Levels			
	10^{-3} PAL[a]	10^{-2} PAL	10^{-1} PAL	1 PAL
260–263	1.3×10^{9}	5.7×10^{-22}	2.58×10^{-69}	1.6×10^{-45}
267–270	3.0×10^{10}	1.67×10^{-14}	2.47×10^{-53}	2.54×10^{-23}
278–282	2.44×10^{12}	4.23×10^{0}	1.93×10^{-18}	3.2×10^{-9}

NOTE: UV photon fluxes at surface (watts m^{-2}).
[a]PAL = present atmospheric level; see text for references.

(1993) concluded that a UV screen would have begun to form at an oxygen concentration of about 10^{-3} of the present atmospheric level (PAL) and that it should have provided a biologically helpful screen by 10^{-2} PAL. For bacteria, UV radiation becomes harmful or produces deleterious effects at a continuous flux of about 400 watts per square meter and at about an order of magnitude less for eukaryotes.

It would appear then that in the absence of organismic or lithospheric screens for ultraviolet radiation an atmospheric oxygen level of 10^{-2} PAL is sufficient to produce a satisfactory UV screen for terrestrial life. But prior to the accumulation of 10^{-2} PAL of oxygen, land migrations of multicellular organisms may not have been possible.

Kasting et al. (1992) and Kasting (1993) have argued that a screen sufficient for the survival of terrestrial life might well have been in place by the early Proterozoic, about 1.5 bya. Holland et al. (1986), basing their argument on paleosols, have proposed that mid-Precambrian pO_2 was on the order of 3×10^{-5} PAL (uncertainty of one order of magnitude) and that by mid-Proterozoic the pO_2 was still "very much smaller than it is today."

2.2.2 Carbon Dioxide and the Greenhouse Effect

The other major environmental change for consideration is the partial pressure of carbon dioxide (pCO_2) in the atmosphere and any subsequent greenhouse effect, or its opposite, excessive cold. Frakes (1979) and Walter (1979) have pointed out that the late Precambrian (0.9–0.6 bya) was marked by a series of extensive glaciations and it seems likely that by that time pCO_2 levels had reached appreciable amounts. If so a major CO_2 greenhouse effect is out of the question and it can be assumed that pCO_2 levels were about the same as they are today.

Overall, the accumulated evidence from models and the lithospheric record indicates major changes in pO_2, pO_3, UV flux, pCO_2, and temperature during the late Proterozoic (1.0–0.6 bya). Although there is no fossil record of land migration during this time, it can be assumed that changing environmental conditions at least began to set the stage for such migrations. Physical conditions before and perhaps at the beginning of this time span acted as a restraint upon migrations of multicellular organisms.

2.3 Geography, Topography, Sediments, and Climates

Since the advent of concepts of plate tectonics an immense amount of study has been devoted to determining the composition, structure, and geographical distributions of the land and oceans throughout earth history. Continuing developments in interpretations are providing increased accuracy and detail, but today at least a broad consensus of approximate conditions throughout the Phanerozoic (from the Cambrian onward) has been reached. Results for earlier eras are more tentative and become increasingly so from the late Proterozoic back into the Archean.

The series of maps in figures 2.2 to 2.8 illustrate in simplified form current understanding of continental positions over the period of time in which initial occupancies of land may have taken place. Added are some of the principal features of the continents, major mountain ranges, and coastal and shelf areas. The spatial relationships of land and ocean clearly impose no constraints on passage from water to land during any time from the Proterozoic onward. Climatic features including temperatures, humidity, rainfall, and winds may have been more critical. Distributions of sediments on the continental margins and in shallow seas provide data on areas where transitions might have occurred and where remains of terrestrial organisms might be found.

Paleomagnetic analyses have supplied information on latitudinal distributions of the land masses. Physical features of the continents,

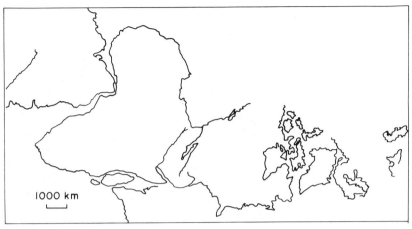

FIGURE 2.2 Sketch map of the continental blocks in the Proterozoic. *After Piper 1977.*

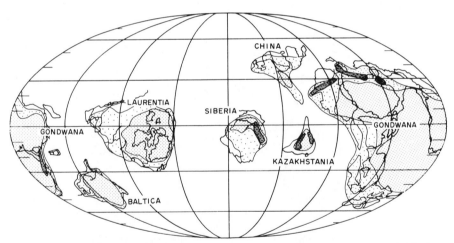

FIGURE 2.3 Paleogeographic map of the late Cambrian. Unshaded,
deep ocean; light shading, shallow seas; intermediate shading, lowlands;
dense shading, mountains. Mollweide projection. *After Scotese et al. 1979
and Zeigler et al. 1979.*

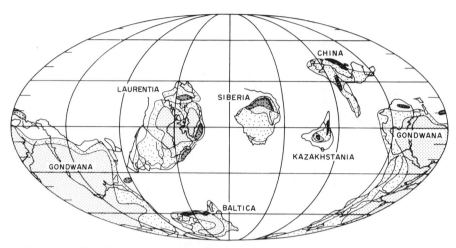

FIGURE   2.4 Paleogeographic map of the middle Ordovician (Llandeilo epoch). Shading as in figure 2.3. Mollweide projection. *After Scotese et al. 1979.*

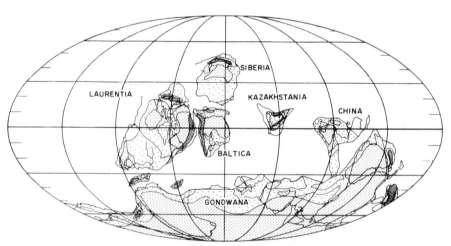

FIGURE   2.5 Paleogeographic map of the middle Silurian (Wenlock epoch). Shading as in figure 2.3. Mollweide projection. *After Scotese et al. 1979.*

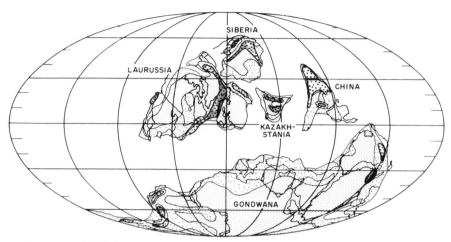

FIGURE 2.6 Paleogeographic map of the early Devonian (Emsian age). Shading as in figure 2.3. Mollweide projection. *After Scotese et al. 1979.*

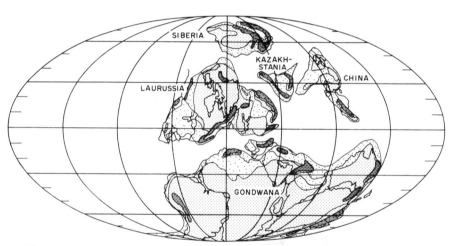

FIGURE 2.7 Paleogeographic map of the early Carboniferous (Visean epoch). Shading as in figure 2.3. Mollweide projection. *After Scotese et al. 1979.*

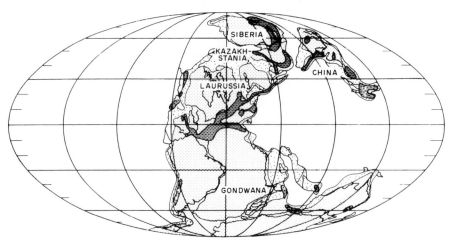

FIGURE 2.8 Paleogeographic map of the middle Permian "Wordian age." Shading as in figure 2.3. Mollweide projection. *After Scotese et al. 1979.*

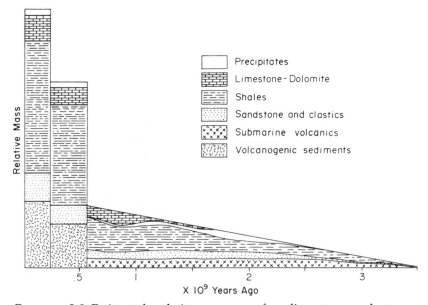

FIGURE 2.9 Estimated relative masses of sedimentary rock types formed over the last 3×10^9 yrs that have been preserved into the present. *Based on Piper 1977.*

TABLE 2.3
*Classification of Environments Relevant to
Invasions of Land*

Oceanic Realm

  Neritic zone

Continental Realm

  Cryptic (infaunal) zone
  Fluvial, including riparian, zone
  Lacustrine, including shore, zone
  Paludal zone
  Terrestrial (epifaunal) zone

Mixed Continental-Oceanic Realm

  Littoral zone
  Supralittoral zone
  Estuarine zone
  Marginal lagoonal zone
  Deltaic zone

their structure, volcanism, and sediments, along with fossils have been important in establishing the more difficult and less certain longitudinal relationships.

Climatological interpretations have been based upon models of atmospheric circulation, derived from meteorological analyses and applied to the globe with considerations of continental positions at particular times. Continental topography and, importantly, interpretations of the significance of distributions of climatologically sensitive sediments (see Robinson 1973), have added confirming information.

A generalized survey of the relative amounts of several categories of sediments deposited through geologic time and that have been preserved into the present is shown in figure 2.9. Marked changes in erosion and deposition occurred, with a major shift at or just before the beginning of the Phanerozoic. Nevertheless, these changes are not sufficiently drastic or sudden to suggest a cut-off time before which shallow water environments and land-water relationships might have been unsuitable for development of terrestrial organisms. Marine-continental interfaces existed back to the time of origin of the continents—probably very early in earth history—and sediments that formed throughout at least the latter part of the Proterozoic indicate the existence of fresh waters into which organisms might have penetrated, and from which some could have migrated onto land. Table 2.3 presents a classification of environments relevant to invasions of the land.

2.4 Changes in Day Length

It is generally believed that year length has remained quite constant through earth history. But there is compelling evidence that lengths of days and lunar months have changed through geologic time. The evidence comes from a number of sources, including geophysical models based on the earth-moon system, tidal friction, continental positions, internal dynamics of the earth, and fossils. Only fossils provide direct evidence of lengths of days and months in geologically ancient times. Day length could be a significant influence in the physiology of organisms, and it might affect capacities of organisms to exist under particular subaerial circumstances.

A consensus exists that day length has increased from the past to the present. The amount of change, however, is far from well delineated. Extrapolations from historical observations and geophysical models have not given a firm basis for interpretations. Fossil evidence, which is more concrete, has come primarily from reef organisms such as corals, along with bivalves and stromatolites—all of which show growth increments of varying widths, thereby making them useful indicators of environmental changes. Much of the attention has been devoted to the Proterozoic (for which stromatolites provide the evidence) and to the Devonian, Carboniferous, and Cretaceous (for which corals and bivalves are key). (See papers in Rosenberg and Runcorn 1975.) The evidence for considerably shorter days in the Proterozoic—600 or more per year (Mohr 1975)—has, however, been largely discounted because of the difficulties in obtaining reliable data.

Corals, based on the analyses and interpretations of Wells (1963, 1970), indicate a year of about 400 days (about 22 hours in a day) during the Devonian. Although this conclusion is now widely accepted, many variables are known to affect growth patterns. Additional studies on corals and bivalves will surely result in more definitive determinations.

For our purposes, however, the uncertainties are not important. The difference between 365 and 400 days, or even a rather wide deviation from either, would have little effect on the physiology of organisms that might pass from water to land. Whether or not the rate of change of day length has been on average constant through time is not established. There is some evidence of irregularities (Rosenberg and Runcorn 1975)—countered, however, by other evidence of generally constant rates. Unless there was a marked deviation from constancy prior to the Devonian, projection of the rate back into the Proterozoic does not suggest that changes in day length would have been an obstacle to transitions of organisms from water to land.

2.5 Survey of General Conditions: Proterozoic to Permian

Interpretations of continental structure and land-sea relationships during the *late Proterozoic* have not yet reached consensus. One position is that stated by Piper: "A supercontinent incorporating African, Laurentian, and probably Baltic-Ukrainian Shields existed without loss of coherence throughout most of this (Proterozoic) time" (Piper 1977:226). Various alternative and modified interpretations can be found in Piper 1983, Hartnady 1991, Dalziel 1992a and 1992b, and Scotese and McKerrow 1990. In this section we generally follow Piper's interpretation.

Figure 2.2 is a map of the Proterozoic continents from Piper (1977). It shows the structure of the Proterozoic supercontinent, "Pangea." (A map from Morel and Irving 1978 gives more information on the position of the continent.) On Pangea, as interpreted by Piper, were extensive grabens which received fluviatile sediments. Massive intrusive activity occurred throughout the Proterozoic, lessening as time passed (Dalziel 1992a, 1992b). During this time the earth's temperature gradients were lessening. How much effect these volcanic and tectonic events may have had on the properties of the atmosphere and ambient temperatures of the land and oceans is, of course, largely speculative.

The late Proterozoic as a whole presents a picture of moderate continental stability, with variation in the rates and amounts of continental erosion and volcanism accompanied by moderate modifications in land-ocean relationships. Rifting was particularly important (Dalziel 1992a, 1992b). Shallow sea environments existed, populated by stromatolites, which decreased in abundance during the later part of the Proterozoic. And it is during this time that sediments (of Vendian age) in several parts of the world first carry remains of complex, largely soft-bodied pelagic and possibly benthic metazoans (see the treatment of the Ediacaran period in Cloud and Glaessner 1982; Fedonkin 1985; Valentine 1989). Among these fossils are presumptive arthropods and annelids, both potential invaders of the land.

It is the *Phanerozoic*, however, that is our prime concern. Interpretations of the continental and oceanic relationships during the Phanerozoic in this and subsequent chapters are based largely on the studies reported in Zeigler et al. 1979 and Scotese et al. 1979. Various modifications of their conclusions have subsequently been advanced, some minor and others more significant with respect to particular groups of organisms and their environments (see the various reports in McKerrow and Scotese 1990). For the most part, however, as far as timing, location, and the nature of transitions from water to land are concerned, these modifications are insignificant.

This statement applies as well to all the paleogeographical maps, figures 2.3 to 2.8. These six maps were selected from a series of forty-two published by Scotese and Zeigler (1979). They have been modified to emphasize continental relief. The Scotese-Zeigler maps provided a basic series from which later interpretations have been made for various times in the Phanerozoic era (see a subsequent series of maps in Scotese and McKerrow 1990). None of these, however, have required modifications that contribute significantly to the central problems we discuss. Because the Scotese-Zeigler maps are widely available and because later changes are primarily devoted to particular detailed analyses (mostly not pertinent to the aims of this book), we have chosen to base our six maps upon the appropriate Scotese-Zeigler maps. Whatever one's terms of reference, however, opportunities for invasions of the continents have existed continuously since the late Proterozoic.

The *Cambrian* period (figure 2.3) lasted about sixty million years, during which time the moderately stable conditions of the Proterozoic came to an end. The Cambrian appears to have begun during a time of extensive rifting. Continental positions and the topography of the land both altered notably. Fragmentation of the Proterozoic supercontinent produced discrete continental blocks. At all times, however, the equator crossed large areas of the continents, implying the existence of equatorial rainy belts. Mountains were rather limited and their influence on global atmospheric circulation probably was minimal. Near the end of the Cambrian continents lay largely in low latitudes, leaving broadly open northern and southern hemisphere oceans.

Continuing the trend established in the Cambrian, the land masses during the *Ordovician* were dispersed largely south of 30 degrees north, with a large, open northern ocean (figure 2.4). Glaciation occurred during this period, as it is known from deposits in both northern and southern Africa. Sea levels dropped and there is ample evidence of active erosion in the continental interiors and margins. Shelf seas were also extensive, providing rich habitats for pelagic and shallow-water benthic invertebrates.

It is in Ludlovian sediments of the late *Silurian* that the first clearly terrestrial organisms have been discovered (figure 2.5). Plant-metazoan relationships appear to have existed on the land by the early Pridoli (Jeram et al. 1990) and were well established by the time of formation of the Devonian Rhynie Chert of the Old Red Sandstone complex in Scotland (Kevan et al. 1975). Global sea level rose during the Silurian with the melting of the Ordovician continental glaciers, and shallow seas spread widely over the margins of the continents.

The massive southern continent, Gondwana, was disposed largely south of 30 degrees south latitude. Laurentia, Siberia, Baltica, and Kazakhstania, except for northern Siberia, all lay within 30 degrees north and south of the paleoequator. The Caledonian fold belt (Morel and Irving 1978) developed as Laurentia and Baltica approached each other, limiting the extent of the long-existent intervening Iapetus Sea. As in the Ordovician, the physical setting during the Silurian seems well suited for transgressions of plants and metazoans onto the land. By the end of the period these invasions of the land were well underway.

The trend of northward movement of the continents and consolidation of their masses, initiated during the Silurian, continued during the *Devonian* (figure 2.6). By early Devonian (Emsian) Laurussia (Baltica + Laurentia) had taken form. The Iapetus Sea was eliminated and extensive mountain ranges developed during the Acadian Orogeny, following the earlier Caledonian episode of mountain building. By Emsian times the "Old Red Continent," which has figured prominently in studies of plants and of freshwater and terrestrial animals of the Devonian, was well established.

Laurussia extended from slightly south of the paleoequator to nearly 50 degrees north latitude, providing a wide range of environments in which land organisms might have developed and flourished. The "Australian" portion of Gondwana likewise lay in favorable latitudes, from near the paleoequator to about 30 degrees south.

The *Carboniferous* (figure 2.7) marks a shift in the focus of studies of terrestrial life, away from searches for the invaders to analyses of subsequent radiations of established terrestrial groups. During the Carboniferous trends continued toward consolidation and development of more extensive land masses in the northern hemisphere, first evident in the late Devonian. This led to near completion of another single supercontinent with final incorporation of China, a complex land mass, occurring in the late Triassic. During the early Carboniferous and persisting into the *Permian* (figure 2.8), lowland coal swamps were broadly developed in the circumequatorial zone, and high latitude glaciation was extensive in the southern hemisphere. Strong fold belts developed (Hercynian, Appalachian, Ouachitan, and Mauritanian), and these contributed to restricted geographic ranges (endemism) of land animals and plants during this long span of time.

The equatorial zones through the Carboniferous and into the early Permian were sites of humid, warm climates and have produced much of the fossil evidence of terrestrial life. With the onset of the early Permian, incipient dryness is found in the equatorial zones, as moist conditions slowly gave way to desert climates. The major zones of continental habitation shifted north and south to middle and upper latitudes, where earlier dry, arid conditions had been quite general (Olson 1985).

2.6 Summary

Terrestrial organisms are preserved in the fossil record for only about one-seventh of the 3.0 to 3.5 billion years of the known history of life on earth. The questions posed by this timing of events have prompted extensive studies of the physical and biological constraints on migrations of plants and animals from water to land.

The first clear evidence of terrestrial organisms comes from sediments of the Ordovician period (475 mya). Records of land and oceanic environments during the last billion years of earth history indicate that occupancy of the land could have occurred far earlier than it actually did. Why, then, the delay? Of all physical factors, the composition of the atmosphere seems the most likely cause. Accumulated evidence from the lithospheric record indicates that major changes in the important atmospheric parameters affecting temperature and UV flux occurred during the late Proterozoic (1.0–0.6 bya), setting the stage for occupancy of land.

Geography, topography, sediments, and climates of the continents have changed markedly during the Phanerozoic. At all times during this era, however, the marine-continental interface, the existence of fresh continental waters, and temperatures over broad areas of the earth appear to have posed no constraints on invasions of the land by aquatic organisms.

3

Plant Transitions to Land

DAVID J. CHAPMAN

The invasions of the land by organisms that are presumed to be multicellular probably began with the plants. The fossil record shows this and it makes ecological sense: land invasions by animals could not succeed until adequate terrestrial food supplies had been established and suitable ecological niches developed.

This chapter describes what is known, and points out much that is still unknown, about plant transitions to land. The focus is the fossil record, but a variety of other lines of evidence deriving from studies of living plants are also explored. The transitions of early aquatic algae onto land can be examined from several angles. Specifically, what were the aquatic ancestors or the migrants? And what were the first land plants and what did they look like? When did this migration occur? How did it happen and how many times did it happen? Where did it occur? We may also speculate about *why* the migration took place. Figure 3.1 summarizes major events in the invasions of land by plants.

The traditional approach to these questions is paleobotanical. It usually involves examining the fossil record, then drawing further inferences from comparisons with existing floras. The fossil record, however, is only a partial record. There is an understandable tendency to compare or to try to relate fossils to extant modern plants. It is easier and tidier to work with that with which we are familiar. But the success of this method depends upon a series of assumptions: that the fossil record can be readily compared to modern plants; that we can identify migratory failures; that we are confident of modern plant phylogeny; and that most early plants have left modern descendants whose relationships we understand. We believe that none of these assumptions is more than partly valid.

Living plants, algal-aquatic, bryophytes, and vascular-terrestrial, provide large amounts of information that supplements information extracted from the fossil record. Major sources of important circumstantial evidence are studies of living plants at multiple levels of complexity: molecular biological, biochemical, morphological, physi-

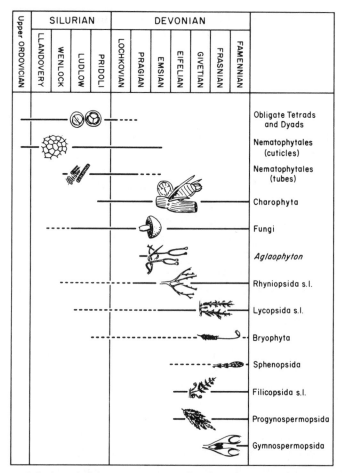

Upper ORDOVICIAN	SILURIAN				DEVONIAN							
	LLANDOVERY	WENLOCK	LUDLOW	PRIDOLI	LOCHKOVIAN	PRAGIAN	EMSIAN	EIFELIAN	GIVETIAN	FRASNIAN	FAMENNIAN	

Obligate Tetrads and Dyads

Nematophytales (cuticles)

Nematophytales (tubes)

Charophyta

Fungi

Aglaophyton

Rhyniopsida s.l.

Lycopsida s.l.

Bryophyta

Sphenopsida

Filicopsida s.l.

Progynospermopsida

Gymnospermopsida

FIGURE 3.1 Approximate sequence of major events in the invasions of the land by plants from Ordovician to Devonian times, as known from the fossil record. Dotted lines indicate aquatic record; solid lines indicate terrestrial record. *After Selden and Edwards 1989.*

ological, and ecological. Current dogma holds that terrestrial plants evolved from algae that resembled and were ancestral to certain modern taxa of algae. There is no serious challenge to this dogma unless one wishes to invoke a hypothetical group of ancestors that has not left any modern relatives and is unrelated to the presumptive migratory group, the green algae. As more data accumulate, particularly from molecular and biochemical fields, this possibility becomes increasingly unlikely.

We believe that an examination from multiple angles of modern terrestrial plants (post-migratory descendants) and aquatic algae (pre-migration descendants) is a valid approach to addressing many of the questions raised above. This method provides a number of possible scenarios or explanations for the land migrations.

Plant migrations from aquatic to terrestrial habitats involved radical changes in environments, which necessitated or brought about major changes in physiology, morphology, and life history. One can infer possible scenarios of what happened and how by comparing these features in modern land plants versus modern aquatic algae. There must be assumptions, but this approach provides the basis for proposing hypotheses to address the major questions. The fossil record and phylogenies deduced from molecular data can then be used to test these hypotheses and to assess the likelihood of multiple migrations and the extent of evolutionary failures along the way. We will end this chapter with a best case proposal for the sequence of events. Table 3.1 presents our preferred classification of plant groups.

3.1 Land Adaptive Features

A comparison of fundamental differences between the two habitats serves as a starting point. The list on p. 5 is illustrative. This list is not inclusive, nor is it without exceptions, but it provides starting points to identify those features that are land adaptive, which could thus be expected only in land plants, and for which we may look in the fossil record to find indicators of the earliest land plants.

By beginning with these differences we can postpone treatment of the complex issues relating to those plants that live in damp habitats—notably, the mosses, liverworts, and hornworts (all of which are usually described as bryophytes). We will not further consider the numerous higher flowering plants (angiosperms) that are secondarily aquatic and that almost certainly returned to the aquatic milieu while retaining some terrestrial attributes. Examples of secondarily aquatic plants are water lilies (*Nuphar, Nymphaea*), freshwater aquatics (*Elodea, Sagittaria, Potamogeton, Myriophyllum*), and sea grasses

TABLE 3.1
Classification of Plants

DIVISION CHLOROPHYTA
  Class Chlorophyceae C, Af
  Class Charophyceae Af
  Class Micromonadophyceae Am, Af
  Class Pleurastrophyceaeae C*, Af
  Class Ulvophyceae Am, Af

DIVISION BRYOPHYTA
  Class Hepaticae C
  Class Anthocerotae C
  Class Musci C

DIVISION RHYNIOPHYTA
  Class Rhyniopsida F
  "Cooksonioids" F

DIVISION ZOSTEROPHYLLOPHYTA
  Class Zosterophyllopsida F

DIVISION TRIMEROPHYTOPHYTA
  Class Trimerophytopsida F

DIVISION LYCOPHYTA
  Class Lycopsida P
  Class Sphenopsida P

DIVISION PSILOPHYTA
  Class Psilopsida P

Af: aquatic freshwater. Am: aquatic marine. C: cryptic. F: fossils only. P: predominantly or fully terrestrial
*Cryptic representatives typically subaerial.

(*Phyllospadix, Zostera, Thalassia*). Other examples of secondary adaptations are plants that live at the marine-terrestrial interface, such as mangroves (*Avicennia, Rhizophora*) and salt marsh plants (*Spartina, Salicornia*).

There are also a few green algae (*Trentepohlia, Cephaleuros, Botryococcus*) that live in "terrestrial" habitats such as moist surfaces on leaves, rocks, or mud. These algae survive, however, without the need for land adaptive morphologies. All, however, are single-celled or few-celled simple thalli.

3.1.1 Desiccation Resistance

An alga of simple morphology (i.e., a filament or thin blade) rapidly dries when removed from the aquatic environment. Algae with more complex morphology and tissue dry more slowly. Prolonged desiccation or excessive water loss may result in death. This phenomenon is readily observed among macrophytic algae living mainly in submerged habitats. Algae lack any effective waterproofing and thus cannot prevent water loss. Some can, however, cope physiologically with water loss. These physiological capacities permit these algae to tolerate desiccation for short periods of time, generally a few hours to a few days. Other morphological adaptations include reduction of the ratio of surface area to volume of the thallus, retention of water within the cavities of the thallus, or increase in cell number for a given surface area. Algae and, like them, mosses and liverworts (bryophytes), cannot actively maintain an internal hydrated environment. Rather, they are *poikilohydric plants* (Raven 1977, 1984, 1985) and survive because of their ability to tolerate dehydration for short periods. We can safely conclude that the early colonizers of the land were small, that they lived in damp environments, and that were poikilohydric.

Water loss over short time spans would be an inevitable challenge, sooner or later, for any plant attempting to survive on land. Time is the enemy. The tolerance adaptations that have evolved in algae living at the water's edge give insufficient protection for permanent habitation in terrestrial or drying, however humid, environments. The migratory lineages evolving into permanent land plants had to develop appropriate countermeasures. If they had failed they would not have been successful at colonizing the land.

There are two principal countermeasures. One involves the development of *parenchymatous organization*—that is, multicellular differentiated tissue (Stewart 1983). Increase in the number of cells for a given surface area of thallus slows the rate of desiccation. If the early land plant inhabited a consistently moist environment—whether by way of rain, dew, high humidities, or damp surfaces—a parenchymatous organization, combined with a small plant form and a physiological ability to tolerate water loss, might have been adequate for survival. This morphological adaptation alone, however, is insufficient for large plants or for those living in environments subject to only short periods of moisture.

The second countermeasure involves the development of a waterproof surface coating. A successful land invasion with subsequent increases in plant size and occupation of mesic and xeric environ-

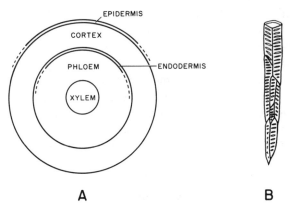

A   **B**

FIGURE 3.2 Diagrammatic cross section of a protostelic vascular plant stem (A) and illustration of a tracheid showing bordered pits (B). *After Gifford and Foster 1987.*

ments required some mechanism to regulate and minimize water loss. In contrast to poikilohydric forms, such plants actively maintain hydrated internal environments. They are thus *homoiohydric plants* (Raven 1977, 1984, 1985; Selden and Edwards 1990). Land plants maintain their internal water environment by the synthesis of a cuticular wax, cutin (Kolattakudy 1980). This hydrophobic polymer provides an impermeable barrier to water.

The aerial parts of terrestrial plants acquire their water and nutrients as solutes from the soil substrate through roots and root hairs. Water and solutes are transported through the xylem (conducting tissue composed of dead cells; figure 3.2) to the aerial parts (leaves and shoots). Water evaporation from the cell surfaces maintains water flow. The water vapor diffuses into the intercellular air space and then out of the leaf. Because cutin (and thus the cuticle) is virtually impermeable to water and also to gases, water evaporation and gas exchange must be regulated by openings in the cuticle of the aerial surfaces. This regulation is achieved in extant land plants by stomata. We can presume that early land plants had some form of cuticular covering, perhaps chemically similar to extant cutin. A waterproof cuticle solved one problem, but created another.

3.1.2 Stomata for Regulating Gas Exchange

While cutin on the epidermal layer minimizes evaporative water losses, unless it is very thin, it also inhibits free exchanges of gases. Free flow diffusion of oxygen and carbon dioxide must be provided for the inner cells to respire and photosynthesize. Such gas exchange may be achieved by pores in the surface of the thallus, or, as is common in extant higher plant leaves, by stomata. Simple pores provide only a rudimentary solution since they cannot vary in diameter. Stomata, however, can regulate to open and close. Stomatal openings remain at a minimum during the night: with no photosynthesis there is no demand for CO_2. During the day, when there is a photosynthetic demand for CO_2, stomata are open. If the water supply becomes inadequate, transpirational water loss and also CO_2 uptake are restricted by closure of the stomata. The opening and closing of the stomata pores are controlled by the two surrounding guard cells (figure 3.3).

The cuticles of modern plants are resistant to both chemical and biological degradation. Those of early land plants were probably also chemically resistant, but we cannot necessarily assume biodegradable resistance. If early waterproof cuticles were resistant to biodegradation, they could be expected to leave fossil records as epidermal sheets of cells, with pores or stomata in the cuticle easily recognizable. These are two fossil markers that can be used as indicators of land colonization and terrestrial adaptation.

The change to a gas phase environment necessitated an additional change in the internal structure of the aerial parts—that is, in the arrangement of cells to provide *intercellular air spaces* (Raven 1984). The addition of intercellular air spaces increased the surface area to volume ratio, thereby reducing for CO_2 the length of the liquid phase diffusion path to sites of enzymatic fixation. Intercellular air spaces are not found in aquatic algae (the morphologically similar spaces in brown algae are not functionally similar). Given the nature of air spaces in "unreinforced" structures such as leaves and the processes of preservation and fossilization, it is probably impossible to ascertain whether or not the small earliest colonizers possessed them. Perhaps one can assume these spaces if a fossil specimen shows evidence of stomata. Raven has rightly pointed out, "It would be

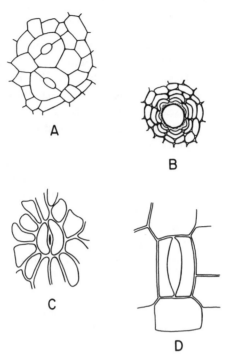

FIGURE   3.3 Pores and stomata of modern plants. A: pore of a liverwort (Conocephalum). B: pore of a hornwort (Anthoceros). C: stomata of moss (Funaria). D: stomata of a gymnosperm. *A, B, C after Scagel, Bandoni, Maze, Rouse, Schofield, and Stein 1982. D after Gifford and Foster 1987.*

interesting (and puzzling!) if stomata occurred in the absence of intercellular gas spaces" (1984:114).

3.1.3 Tissues for Water Conduction and Mechanical Support

Terrestrial habitats are water and nutrient deficient, at least for the aerial parts of plants. Cutin restricts evaporative water loss but also water infusions from moisture depositing on plant surfaces. Water and nutrients must thus come from the soil. These nutrient and water deficits are remedied by subterranean absorption structures, roots and root hairs. Movement of water and nutrients to the aerial and photosynthetic tissues is achieved through conducting or vascular tissue, principally *xylem*. The cell walls of the conducting cells of xylem (*tracheids*) are characterized by thickenings composed of lignin, which is a polymer of C_6-C_3 phenylpropanoid units. The water resistant and bioresistant properties of lignin make it an excellent structural material for water/solute conducting cells.

The earliest, presumably very small, land plants might have been able to survive without lignified water-conducting cells. However, further evolutionary advances in plant size and upright posture necessarily required the development of water-conducting tissues. The structure and function of modern plants suggests there is no alternative for survival in dry environments. Terrestrial plants taller than a meter must function as homoiohydric plants with concomitant gas exchange features (Raven 1984).

The xylem performs dual roles of fluid conduction and providing structural integrity. While increased height confers potential evolutionary advantage, in terms of access for photosynthetic parts to light in a canopy, it also necessitates increased mechanical support. Bundles of lignified cells form rigid tissues which provide compression-resistant mechanical support. These are rigid enough to support aerial structures, but flexible enough to permit bending, twisting, and stretching. The stem or trunk represents the obvious and extreme development.

Algae do not synthesize lignin. Mosses do synthesize phenylpropanoid (C_6-C_3) dimers related in composition to lignin polymers, but they do not make true lignin. Although mosses lack lignified cells, they do possess fluid-conducting cells called leptoids and hydroids. The importance of lignin as a mechanical support is demonstrated, in a backhanded fashion, by aquatic angiosperms. Secondarily aquatic plants show greatly reduced lignification.

The earliest migratory plants almost certainly did not need the structural support required by later upright plants. They were probably prostrate forms, and what short erect parts they might have had could have been supported by existing multicellular structures. Multicellularity, cell turgor, and parenchymatous organization provide sufficient structural reinforcement for small erect plants (Speck and Vogellehner 1988). The best mechanical support is provided by strong but flexible cell walls. While a number of plant structural polysaccharides (e.g., celluloses, xylans and mannans of macroalgae) do provide some support when formed into thick walls and com-

bined in parenchymatous tissues, none match lignin for this purpose. Using the psilophyte *Psilotum* as a model, Niklas (1990) has
shown that lignified xylem is the principal means for aerial support.
    Because xylem provides for the transport to the aerial photosyn-
thetic tissue of water and nutrients absorbed by the subterranean
roots and root hairs, there must in turn be some pathway for transport of organic photosynthate from the leaves to the nonphotosyn-
thetic, actively growing regions. *Phloem* serves this function. Phloem
tissue is composed of living cells and includes sieve tubes, which are
vertical rows of elongated cells. These sieve tubes are the conducting
elements of the phloem. Lignification is absent in phloem. While
phloem is a necessary attribute of an aerial system, and thus a terrestrial adaptation, its cellular nature does not lend itself to preservation, except as part of a larger, completely preserved section of the
plant.

3.1.4 Spores

The migration to land also involved major changes in reproductive
strategies. In algae, reproductive cells—whether gametes or zoospores—are frequently motile (achieved by means of flagella). With
a few notable exceptions, there is always at least one stage of the life
history that produces motile reproductive cells. These exceptions
include red algae, the conjugating green algae (Conjugales) in the
Charophyceae, certain orders of green algae (e.g., Chlorococcales),
and the pennate diatoms. Both male and female gametes produced
by the gametophyte may be motile, or just the male as in oogamous
reproduction where the female cell is a non-motile egg. The sporophyte, depending upon the taxon, may (i.e., zoospores) or may not
(i.e., aplanospores, tetraspores, monospores) produce motile cells.
In those classes of algae composed principally of flagellated members—such as Prymnesiophyceae (golden brown algae), Dinophy-
ceae (dinoflagellates), and Rhaphidophyceae (yellow-green algae)—
the vegetative phase itself is motile. Motility is a perfect attribute for
reproduction in an aqueous environment, particularly if it is combined with an ability to detect and respond to chemical signals or
directional light beams and irradiance gradients.
    Reproductive systems independent of water were essential for
evolving land plants. All extant gymnosperms and angiosperms

(even those that secondarily have returned to water environments) have such systems. Regardless of the vector (e.g., wind—or later, insects), land plant reproductive cells must be adapted to water-deficient environments; they must also be resistant to possible mechanical damage. These requirements were met by the development of water-resistant, reinforced walls. The polymer coat, sporopollenin, found in the walls of many extant spores serves both these purposes.

3.2 Four Footprints of the Transition

Comparisons of modern terrestrial plants with aquatic ones suggest a number of major changes that must have occurred during the land migration. These in turn suggest some of the events that could have taken place and answer some of the questions that deal with "how" and "what." Fossils showing evidence of these changes are indicative of the actual land migration. They help establish possible time frames and biogeographical locations for the migration.

There are four major adaptations that we regard as the "footprints" of the transition: (1) the lignified tracheids of water-conducting xylem, (2) protected reproductive cells, (3) cuticular pores and stomata, and (4) cuticle itself. What are the chances of fossilization of each of these innovations?

A bioresistant and chemically resistant cuticle could be expected to be preserved during diagenesis. Should that preserved cuticle show evidence of pores or structures corresponding to stomata, then the fossil record is even more useful. The same can be said with regard to water-conducting tissues. Modern lignin, a polyphenolic polymer forming part of the chemical basis of wood, is very resistant to both chemical attack and to degradation by biological agents (a major exception being some basidiomycetous fungi). Preservation of lignified plant parts would thus be favored. If lignin served as the basis for both upright mechanical support and xylem walls early in the transition, the fossil record of that period surely would have recorded the innovation. Protected or resistant reproductive cells should likewise lend themselves to preservation. Pollen grains, with their thick and resistant walls are well documented in the paleobotanical record.

Fossilized tracheids can be considered conclusive evidence for the presence of vascular land plants. No extant algae has water-conducting tissues; it is far from parsimonious to entertain the possibility that the algae lineage ever took this step. Cuticular pores and stomata, which if found in the fossil record can be assumed to be parts of gas exchange mechanisms and homoiohydry, are very good presumptive evidence for the occurrence of land colonization. They are known to occur in the extant flora only in bryophytes and vascular plants.

Recognizing ancient plant fossils is not always easy. Tracheids, stomata, spore tetrads and spores with trilete marks are all typically nonalgal "plant." They have no counterparts in either algae or animal phyla that could be mistaken for plant remains. Cuticle, however, is also a feature of arthropods. Gensell et al. (1990) have pointed out that under some circumstances conclusive distinctions may not be possible. Thin sheets of monostromatic algae (e.g., green, brown, or red algae) with thick cell walls might also leave cuticle-like remains. Fossil cuticle without pores or stomata is only a possible indicator of plants and can easily be misinterpreted.

In order to identify ambiguous fossils one may have to rely on other criteria, such as paleochemical signatures. But even this approach is suspect because its baseline data are obtained from modern organisms. There are enough enigmatic fossil remains to suggest that many algae or perhaps early land plants have left no close living relatives and were evolutionary failures. The array of living plants provides only some help. If one looks at the living presumed descendants of the most likely ancestral alga and of the presumptive primitive land plant, there is a wide gap that remains to be bridged.

With few exceptions all photosynthetic plants possess a gametophyte generation. If this is the only generation in the life history (i.e., there is no alternation of generations) then there is no problem of assignment of reproductive cells to sexual status. If the gametophyte alternates with a sporophyte, then a resistant reproductive cell could be derived theoretically from a sexual fusion (e.g., zygote or hypnozygote) or by way of asexual production (spore). The exigencies of sexual reproduction involving cell fusion virtually preclude thick resistant walls on gametic cells. Gametes of land plants can be expected to occur and fuse with the egg only in highly "protected" microenvironments.

On the other hand, there are definite evolutionary advantages for plants having reproductive cells that can withstand the stress and

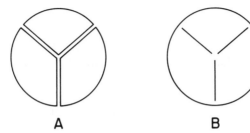

FIGURE 3.4 Diagrams of a spore tetrad (A) and a spore with trilete mark (B).

potential damage of being wind borne. We argue that any such cells must be pollen (microspores bearing the male gametophyte) or spores, originating from sporophytes. Spores normally are formed meiotically in the life cycle. Prior to their release they may have a tetrad arrangement, which leaves trilete marks (figure 3.4) on each of the four spores. The tetrad form is commonly found today in lower vascular plants and it is also known from the bryophytes.

Meiotic tetraspores are known from extant algae (e.g., red algae and Dictyotales brown algae), but there are no recorded instances of thick or reinforced, walled tetraspores or tetrads that would lend themselves to fossilization. This is to be expected in aquatic environments. Algal spores that might be fossilized are invariably single, usually calcified or silicified, or occasionally with thick resistant walls. Such spores (e.g., hypnospores), however, are more often a response to severe or adverse conditions and are not part of the normal gametophyte/sporophyte life history. Equally important is the absence of trilete marks on algal spores (see also Gray and Boucot 1977a).

It is important to remember that trilete marks are observed only on liberated spores. Tetrads may be observed in place prior to shedding. In the absence of preserved tetrads with resistant walls the presence of sporangia per se is not a presumptive footprint for land adaptedness. However, it may be argued that sporangia with thick, resistant, and preservable walls do represent land-adapted features.

Fossilized tetrads and trilete spores can be regarded as presumptive evidence for land plants, although some authorities (Banks 1975a, 1975b; Chaloner 1985) have argued that in the absence of other more persuasive evidence, they should not be taken as indicative of either a colonizing or established land flora. If one accepts this latter view, one must explain what these features in fact are, because there is no good basis for assuming they are of algal origin.

Modern land plants have provided some insight into what could have happened during the transition. It does not follow that any or all of these adaptive innovations occurred at the time of migration, that they all appeared in the same taxa, or that they occurred simultaneously. We thus must turn to the fossil record for further evidence.

3.3 Early Land Plants

The earliest plants in the fossil record that are widely viewed as terrestrial lived in the late Silurian and early Devonian. These have been placed in the extinct class Rhyniopsida (within the phylum Rhyniophyta), and they are sometimes called, simply, "rhyniophytoid-like" plants.

The earliest macroscopic organisms known that are presumably photosynthetic or "plant-like" are much older. These include *Rhabdoporella* in the Cambrian and *Grypania* far earlier—about 2.2 bya in the Precambrian (Han and Runnegar 1992). In all early examples there is no evidence to suggest that they were terrestrial or associated with colonization of the land. Rather, they are regarded as perhaps marine algae, or at least analogous to marine algae.

3.3.1 The Rhyniopsida

The Rhyniopsida were small branched plants, with some (apparent) vascularization, parenchymatous organization, and terminal sporangia, but they lacked any leaflike structures. The presence and degree of vascularization, however, are still in question.

Rhynia gwynne-vaughanii (figure 3.5) is the type for this class (Kidston and Laing 1917, 1921), discovered in the Rhynie Chert of the Old Red Sandstone of South Wales. The presumed age of these beds of the early Devonian is 390 mya. *Rhynia* demonstrates tetrad spores, tracheids, cuticle, and stomatal pores. Stomata appear on the stem of this leafless plant. The vascular arrangement is protostelic (see figure 3.2), meaning that solid xylem is surrounded by phloem. This organization is regarded as primitive (Gifford and Foster 1989). It is also observed in extant lower vascular plants.

Since the discovery of *Rhynia* a number of other presumptive

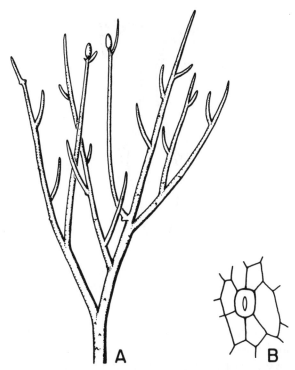

FIGURE 3.5 Reconstruction of *Rhynia gwynne-vaughanii* showing whole plant (A) and stomata (B). *After Edwards 1980.*

rhyniopsid plants have been discovered. They, too, have the basic simple/primitive features of the class, as represented by *Rhynia*. There is disagreement, however, about which of these other plants are true Rhyniopsida. Early workers regarded all the simple fossil plants (*Rhynia, Cooksonia, Horneophyton, Renalia, Taeniocrada, Steganotheca, Hicklingia*) as members of the Rhyniopsida. Taylor (1988) and Edwards and Edwards (1986) have argued that this class cannot be all inclusive. The latter workers have proposed restricting the class to those genera that have smooth axes, terminal sporangia, and xylem (*Rhynia, Renalia, Taeniocrada*). Other genera—notably, *Cooksonia* (figure 3.6) and *Steganotheca*—are designated simply as "rhyniophytoid" or "cooksonioid." These distinctions have important bearing on early plant phylogeny.

The more discoveries in the fossil record, the more complicated the deductions of the phylogeny of early land plants become. How-

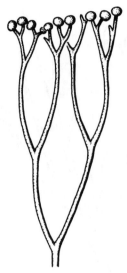

FIGURE 3.6   Reconstruction of *Cooksonia caledonica*. *After Edwards 1970.*

ever, this same record—showing spores, cuticle, stomata and pores, tracheids, and thickened walls—adds credence to the physiological arguments we presented earlier as to the evolutionary innovations that must have attended the migration.

There are only a few genera within each plant lineage that was making the transition to which one can unequivocally attribute two or more of the four "footprints." *Hsüa* (middle Devonian) demonstrates sporangia, trilete spores, stomata, and vascular thickenings. *Taeniocrada* (Devonian) possessed sporangia and xylem. Trilete spores, sporangia, and thickened tracheids have been observed in *Renalia* (early Devonian). Overall, in these earliest plants stomata and pores are observed less frequently than are the other three features we regard as diagnostic of terrestrial adaptations. This may reflect poor preservation and a poor record, rather than a temporally significant absence.

3.3.2 Cooksonia and Other Rhyniophytoids

The rhyniophytoids are simpler, often lacking the full complement of land adaptive morphology found in the Rhyniopsida. The absence of tracheids or thickenings could be significant; Edwards and Edwards "wish to emphasize the probability that early in the colonization of the land there were a number of 'evolutionary' experiments involving both the structure and functioning of water conducting tubes" (1986:216). However, one must ask if such absences are not just artifacts of an incomplete fossil record. Recent discoveries suggest that this may be the case.

If one accepts this separation, then the Rhyniopsida represented in the late Silurian and into the Devonian by such fossils as *Rhynia* and *Renalia* were clearly fully adapted land plants. The rhyniophytoids, extending back to the mid-Silurian, may not have been true land plants but instead inhabitants of transitional environments. This possibility has been raised by Edwards et al. (1986) and Gray (1985a,b). The oldest *Cooksonia* specimens appear to lack vascular strands, suggesting that each cell within these small plants was able to obtain water directly from the environment. Stomata appear to be absent. The thin-walled tissues suggest direct gaseous diffusion, or even diffusion from a surrounding aqueous milieu. On the other hand, other *Cooksonia* fossils seem to indicate stomata (Edwards et al. 1986) and vascular strands (Edwards et al. 1986, 1992).

One is tempted to look for migratory plants that were adapted to very damp habitats or even occasional periods of submersion. However, there are problems. The rock matrix in which *Cooksonia* fossils are found does not reveal unambiguously the environmental conditions. Moreover, one cannot know whether the migration began with single- and few-celled algae or with complex multicellular algae. We can say that some of the late Silurian fossils possessed some of the land adaptive features previously discussed. There were probably many evolutionary experiments at that time in the handling of water insufficiency. The true Rhyniopsida, clearly land adapted, were products of just one set of those experiments.

The only other rhyniophytoid to match *Cooksonia* in age is *Steganotheca*. Kovacs-Endrody (1986) claims that *Promissum pulchrum*, described from the earliest Silurian, is a vascular plant. However, none of the four footprints—xylem, protected reproductive cells, cuticle, and pores or stomata—can be identified in the figures of *P. pulchrum* that Kovacs-Endrody provides. Rayner (1989) has even been suggested that *P. pulchrum* is not a plant.

Another rhyniophytoid of potential significance is *Aglaophyton*.

This was originally described as *Rhynia major*. Reexamination of original material and new specimens has led to the suggestion that the cells in the center of the conducting strand more closely resemble the leptoids and hydroids of bryophytes than the vascular tissue tracheids (D. S. Edwards 1986). This should not be interpreted as placing *Aglaophyton* intermediate between bryophytes and vascular plants. The importance lies in helping deduce the evolution of function as it pertained to the migration.

Despite the disagreements on taxonomic delineation of the Rhyniopsida—which fossil specimens are truly part of this class and which simply resemble the class and are therefore rhyniophytoid—there is consensus that the very early land flora possessed features common to both. These plants appeared during the late Silurian, flourished through the early Devonian, and disappeared at the beginning of the middle Devonian.

3.3.3 Is *Psilotum* a Good Model?

The early part of this chapter examined some of the changes that might have been needed for the actual land migration. The fossil record largely substantiates that these changes occurred. We now ask: what is the closest extant relative to this early vascular land flora?

The obvious comparison is with the psilophyte *Psilotum*, often regarded as the simplest of vascular land plants alive today. *Psilotum* may be simple, but does that make it primitive from a taxonomic perspective?

This genus shows a sporophyte stage with simple upright dichotomous branching, underground rhizomes, simple vascular structure, and homosporous sporangia produced at the terminal ends of short axes. The gametophyte is a small, subterranean plant covered with rhizoids. The genus has traditionally been used as a guidepost to the morphology of the earliest vascular plants because it looks so much like the fossil remains of the long-extinct Rhyniopsida. This interpretation (Gensell 1977; Rouffa 1978) is not, however, problem-free (Stewart 1983). The rhyniopsids disappeared in the middle Devonian and there is no subsequent fossil record, as one would expect of any *Psilotum*-like relatives.

The alternative explanation (Bierhorst 1971; Gensell and Andrews

1984) shows *Psilotum* arising out of the ferns. Until recently the argument was unresolved. Recent molecular analysis of the chloroplast genome, with reference to the gene order (Raubeson and Jansen 1992), does place *Psilotum* closest to the ferns. The fossil record of ferns suggests that the first fern probably evolved in the Carboniferous or Permian. If this phylogenetic placement is substantiated by other independent data (c.f. Chapman and Buchheim 1992), then *Psilotum*, though simple in morphology, can no longer be regarded as primitive—it could not have arisen before the ferns evolved. If this is the case, where then do the Rhyniopsida fit in the evolutionary sequence?

Whatever the answer, the Rhyniopsida must be regarded as the simplest known vascular land plants. They are also frequently held to be the most primitive; a view colored perhaps by the *Psilotum* influence but also supported by their position as the earliest recorded fossil megaphytes.

Among living plants the lycopods, sometimes called club mosses and which include the genus *Selaginella*, would now appear to be the most primitive vascular plant—or at least the group whose ancestry can be traced back the furthest. In an attempt to close the gap between the morphologically primitive modern land plants and the proposed earliest land plants, we can ask to which taxa (if any) did the Rhyniopsida give rise, or, more appropriately, which later taxa share the closest common ancestor with this extinct class? The record suggests two more recent, but also now extinct, groups were most closely related to the Rhyniopsida. These are the Zosterophyllopsida and Trimerophytopsida.

3.3.4 Zosterophyllopsida and Trimerophytopsida

The classes or subdivisions known as Zosterophyllopsida and Trimerophytopsida have a basic morphology similar to that of the Rhyniopsida. Whereas the Rhyniopsida were dichotomously branched, smooth-stemmed plants with single terminal sporangia, the Zosterophyllopsida had lateral sporangia on dichotomous or pseudomonopodial shoot stems that were either smooth or possessed simple (non-leaflike) protuberances. The Trimerophytopsida are distinguished from the Zosterophyllopsida by the clusters of sporangia on divided fertile branches. Like the Rhyniopsida, both of these other

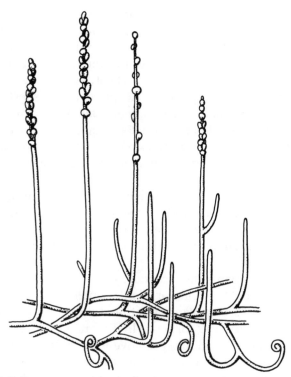

FIGURE 3.7 Reconstruction of *Zosterophyllum myretonianum*. *After Gensell and Andrews 1984.*

classes possess the simple protostelic xylem arrangement, exarchic in the Zosterophyllopsida but centrarchic in the other two.

The Zosterophyllopsida, well documented throughout the Devonian, has a fossil record that slightly precedes that of the Trimerophytopsida. Two examples well illustrate this group. *Zosterophyllum* (figure 3.7), known for nearly a hundred years, has been recovered in Scotland and Wales, Belgium, New Brunswick (Canada), and China. *Gosslingia*, known from a short time span in the early Devonian, has a record through Wales, Germany, Russia, and North America. These two genera are smooth-stemmed forms, and it is easy to imagine how they could have evolved from the Rhyniopsida. There are also a number of genera (e.g., *Crenaticaulis*) with blunt projections on the stem, while others (e.g., *Kaulangiophyton, Sawdonia*) have sharper

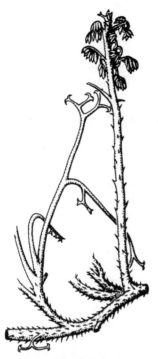

FIGURE 3.8 Reconstruction of *Psilophyton crenulatum. After Doran 1980.*

spinelike projections. (For a more detailed discussion of these and other genera see Stewart 1983 and Gensell and Andrews 1984.)

Zosterophyllopsida comprises genera that are unquestionable land plants but also that are clearly more advanced than the Rhyniopsida. Besides the critical features of vascular tissue and sporangia, some of the remains have been identified as having stomata on the stem. This indication of gas exchange is expected of terrestrial plants. As would be expected in the absence of leaf structures, stomatal features represent strong presumptive evidence for epidermal and subepidermal photosynthetic tissue.

The Trimerophytopsida, known from the early to middle Devonian, are distinguished principally by the occurrence of sporangia in clusters on definitive fertile branches (as opposed to vegetative ones). Although it is not the type genus, *Psilophyton* serves as an excellent descriptive representative. A number of species in this genus have been described from the early to middle Devonian, principally from the Gaspé Peninsula in Quebec. There is considerable variation and complexity in size and branching, but a particularly well-preserved sample of *Psilophyton crenulatum* from the lower Devonian (Doran 1980) demonstrates the habit (figure 3.8).

3.4 The Sequence and Time of Events

In this section we explore when the land migration likely began and the timing and sequence of the various stages. We believe that a good way to establish a time for the onset of migration is to look for the appearance in the fossil record of the first land adaptive morphology. We propose to further address the issue by taking what we call the four footprints of the transition (xylem, protected reproductive cells, cuticle, and pores or stomata) and hypothesizing a plausible sequence for their appearance. How well does the fossil record fit with this sequence? What are the deduced times for the beginning of the migration and the establishment of true land-adapted plants?

3.4.1 A Proposed Sequence of Events

We can safely assume that the early land plants evolving by way of transitions from aqueous, through damp or semiaquatic, to dry and water deficient environments did not simultaneously acquire all of the key attributes we associate with fully terrestrial habitation. In deducing the probable sequence of acquisitions we accept the idea that the land flora arose from green algal clusters. In the absence of persuasive arguments to the contrary we further speculate that the earliest migrators, not totally preadapted for terrestrial environments, were small prostrate thalli inhabiting damp surfaces.

Our proposed sequence has spores and thin cuticle or simple water-resistant coating evolving first, followed by xylem and then stomata. This scenario does not temporally separate spores and cuticle. Raven (1985) proposed a similar sequence, but argued for spores first. Stomata are placed last in both his and our sequences; they probably evolved from simple pores that appeared at the same time as xylem. It is true that our sequence is influenced both by what the fossil record indicates and by morphological resemblances of extant forms of green algae to the fossil record of early plants. However, we believe that our sequence could be deduced independent of any fossil record; the fossil record, rather, is an important test of (and support for) our proposed sequence.

Spores adapted for nonaqueous dissemination were perhaps the first adaptive morphology of the earliest plants to venture onto land. Before spores evolved, plants would have been able to reproduce only when submerged or in wet microenvironments. We assume that the migration started with plants in damp environments. Water-proofing and gas exchange, water and nutrient transport systems were not essential. However, reproductive structures adapted to nonaqueous distribution would have been extremely advantageous for coping with the patchiness of favorable environments.

We also feel confident in assuming that the early migrators were small, probably erect, or perhaps even partially prostrate. If this was the case, one could expect cuticle as the next adaptation. Arguably there was no need for water-conducting tissue.

These arguments could be taken to imply discrete, delineated steps. However, we believe it is likely that cuticle first evolved virtu-ally contemporaneous with the evolution of nonaqueous spore repro-duction. Further evolution would likely have next brought about increased size and probably height. Such plants could still survive without a complex (xylem) system for water transport. However, pores for gas exchange would be needed once a thick, highly water-resistant cuticle had developed. Once pores evolved to stomata, and early thickened cell walls to tracheids, the adaptive morphologies for a true land plant would then be in place.

We offer this proposed sequence with one extremely important cautionary note: *there is no a priori reason to assume that these sequences occurred only once.* Separate migrations, failed or successful, could well have resulted in different sequences at different places and times.

Our proposed sequence has been drawn exclusively from reason-ing based on functional needs. But does it corroborate with the fossil record? The next three subsections will explore the fossil evidence, beginning with the feature that appears to be the most recent (tra-cheids) and concluding with the feature that appears to be the most ancient (spores).

3.4.2 The Earliest Tracheids

The critical overall span of time with which we are concerned involves about fifty million years, from the mid-Ordovician through the Silurian. It is easiest to work backward from the first identifiable land plants with all the adaptive attributes. *Cooksonia* (figure 3.6), discovered by Lang (1937) in Welsh beds of early Devonian age, can serve as the benchmark. Various specimens of *Cooksonia* have spores and sporangia, some have vascular tissue, and some clearly have stomata. Rhyniopsids, appearing a little later, have similar and more complete sets of these key adaptive features.

The simple structures of *Cooksonia* (which seem to have grown no taller than a few centimeters) are readily apparent. There are no microphylls or lateral branches but instead simple dichotomous branches with terminal sporangia. *Cooksonia*-type sporangia have also been recovered from the earlier Wenlockian (Edwards and Feehan 1980). The early specimens from the Pridolian and Wenlockian appear to lack vascular tissue (Edwards et al. 1983), whereas early Devonian *Cooksonia* were vascularized (Edwards et al. 1992).

The various *Cooksonia* fossils, and the Rhyniopsid flora of the early Devonian, point to an established, but simple, land flora by about 425 mya. Further consideration of the individual adaptive features permit us to identify the earlier time period for the beginning of the land migration.

While spores, cuticles, and stomata (and pores) are evidence for land invasions and of early presumptive land plants, the record before Cooksonia currently does not distinguish between bryophytes and nonvascular precursors to the vascular plants. The distinction is important if, as we believe, the bryophytes represent a separate evolutionary lineage deriving from early partially terrestrial plants. In other words, vascular plants did not evolve from the bryophytes that may have already been well established in terrestrial environments. Rather, vascular plants arose from an altogether separate lineage venturing forth from water.

The distinguishing evidence for vascular plants must be provided by presumably lignified tubes or tracheids with wall thickenings, and by megafossils whose form appears by reasonable assumption and conjecture to be that of a typical primitive land plant. (The hydroids of some bryophytes are elongated water conducting cells. While these have thickened walls, their walls are not lignified). A structure that can be identified unambiguously or with at least near certainty as a tracheid can be regarded as evidence for a land plant. However, that entity will be a *vascular* land plant. As in bryophytes, land plants (at least damp environment inhabitants) of simple and small size might not need vascular tissue. Thus tracheids will not indicate when the land migration actually occurred but rather when erect plants with a need to adapt to the water deficiency stresses of terrestrial habitats had started to establish themselves.

Banded tubular elements and thickened tubes have been recovered from the early Silurian (Pratt et al. 1978; Niklas and Smocovitis 1983) and from the late Silurian (Edwards and Davies 1976). Other fossil elements, lacking thickenings and tracheid-like structures have also been identified from the late Silurian and Devonian. Although the exact function of these latter elements is unknown, it may be assumed that they played a role in water conduction.

If banded and thickened cells are similar to tracheids of extant plants, a lignin component may well have been present in the walls. Putative lignin derivatives—i.e., simple phenolic compounds—have been identified in remains from the early Silurian (Niklas and Pratt 1980). However, we must be careful that enthusiasm does not drive the creation of proof. Simple phenolic compounds do not prove lignin and do not prove vascular plants. Bryophytes, although not synthesizing lignin, are known to produce lignols (Erickson and Miksche 1974; Miksche and Yasuda 1978) and lignin-type derivatives (Logan and Thomas 1985), both of which could leave simple phenolics in the fossil record. Thus simple phenolics serve only as corroborative evidence, the aggregate of which suggests that the flora of the Silurian had developed conduction tissues to deal with problems of water deficit. This in turn supports both an established colonization of the land by the Silurian and the existence of an adaptive morphology, specific for the problems of true terrestrial habitats, by the early Silurian, perhaps 435–430 mya.

3.4.3 The Earliest Stomata and Cuticles

Gas exchange structures would be strong evidence for an established terrestrial flora. They appear in the early Devonian with *Zosterophyllum* and *Orestovia*. The earliest identified stomata come from the axis of an early Devonian *Cooksonia* (Edwards et al. 1986), suggesting that the erect aerial axes were involved in gas exchange. Does the absence of vascularization in some *Cooksonia* (known also from the middle Silurian) suggest that stomata—and presumably internal air spaces—were an adaptation that preceded vascular tissues? This would be the reverse of the sequence proposed by Raven (1985) and by us.

The earliest cuticular remains have been recovered from the late Ordovician (Gray et al. 1982) and the early Silurian (Edwards 1982; Al-Ameri 1984). These small fragments lack demonstrable pores, strengthened holes, or stomata—thus complicating a determination of their origin—although the lack of such may reflect only a lack of discovery and not actual absence. Presumably these cuticular fragments derive from early land plants, although a possible aquatic (= algal?) source cannot be ruled out. Neither does this rule out the possibility of plants that are semiaquatic, spending only partial time emersed when waterproofing would be critical. There is some evidence from paleochemistry that these cuticles may in fact be chemically similar to higher plant cuticle. An optimistic interpretation (with which we agree) that these cuticular fragments are indeed remains of higher plants would then place the beginnings of the vascular land flora at the late Ordovician or early Silurian. This would put colonization at about 440 mya—which is 10–15 million years earlier than the date derived from tracheid fossils.

3.4.4 The Earliest Spores

Determining the time of appearance of tetrad spores poses a dilemma. The terrestrial environment and the advantage of asexual sporophytic generations with spore dispersal systems independent of aqueous environments would argue for the early evolutionary acquisition of spores. However, modern representatives of the putative ancestral assemblage of green algae lack a sporophyte generation—there is no alternation of generations. Terrestrially adapting plants must have freshly evolved a life history with alternation of generations, and they must have done so almost immediately. If the earliest aquatic/terrestrial plants possessed only a gametophyte generation reproducing in a water film, then spores (and trilete marks, in particular) would be poor temporal indicators of the earliest stages of the transition. If those early pioneers did in fact lack a sporophyte stage, then one could expect very limited spatial dispersal of new generations. There are thus distinct advantages to the evolution of the sporophyte generation. We regard spores (and particularly tetrad spores) as legitimate indicators of the earliest phase of the transition.

The evidence then for the earliest presumptive land plants comes from formations in Libya and South Africa. (The claims for a Cambrian vascular plant, *Aldanophyton antiquissimum*, from Siberia [Kryshtofovich 1953] have not been substantiated; it is very difficult to ascertain any plantlike affinities from the photographs of the impressions.) Rocks of late Ordovician (Caradoc) age in Libya have provided spore tetrads and cuticular sheets (Gray et al. 1982). Spore tetrads of that time period have also been found in Ghana, Czechoslovakia, and the United States (reviewed by Gray 1985a). More recent investigations in South Africa (which may have involved strata that were very early Silurian; Gray et al. 1986) and North Greenland (Nohr-Hansen and Koppelhaus 1988) have also identified spore tetrads. By the early Silurian (summarized in Gray 1985a,b, 1988; Taylor 1988), single trilete spores appeared; while we argue here that such spores are not algal in origin, Johnson (1985) has proposed that they have an algal origin. This is hard to accept when one considers the nature of extant algal tetraspores and the unlikelihood of their preservation.

Gray (1985a,b) and Gray et al. (1986) have argued that the earliest tetrads in the fossil record (mid-Ordovician to early Silurian) were produced by liverworts, "proto-liverworts," or some plant "morphologically but not necessarily phylogenetically intermediate between green algae and vascular" plants. Spores recovered from the late Llandoverian to Pridolian (Silurian) and which include trilete forms have been considered to signal the presence of vascular plants. However, the difference between tetrads of bryophytes and those of some vascular plants is not great. The more conservative interpretation would be that they are not algal. If one then disregards the remote and unlikely possibility that these tetrads are algal spores, one cannot escape the conclusion that "the first adaptive radiation of plants onto the land occurred in the mid-Ordovician" (Gray 1985a:187). This suggests that a land migration was in place by about 450 mya, and perhaps earlier.

Evidence from spores thus puts the land transition at a time at least ten million years earlier than that indicated by stomata and cuticle—and at least twenty million years earlier than that indicated by tracheids. In summary, we can place a vascular land flora (fully terrestrial) as well established by 420 mya, with the very beginnings of the transition occurring at 450–460 mya. The critical time is therefore a span of approximately 40 million years from the mid-Ordovician to mid-Silurian.

3.5 Gametophytes of Early Land Plants

In the previous sections we have touched on three critical issues that were not fully discussed. These are the question of gametophytes, the possibility of bryophytes as early migrators and as progenitors of vascular land plants, and the phylogeny of all of the plants involved.

3.5.1 Where Are the Gametophytes?

The modern land flora, with dominant sporophyte generations, suggests that the rise to dominance of the sporophyte was an important feature of the land plant invasion and subsequent radiation. How did this occur?

Many extant taxa of algae have life histories that suggest that their own ancestors could have given rise to sporophyte-dominant life cycles. The overwhelming abundance of probable and definite sporophytes among the earliest land plant fossils reinforces the idea that the domination by the sporophyte in modern land plants is in fact a good model for the changes during the invasion. Because modern land plants have gametophytes that range from small, but independent, soil surface thalli in ferns and lycopods to further reduced and dependent forms in flowering plants, we must ask what was the form of the gametophyte in the earliest land plants. A survey of extant plants that are probable descendants of those groups with which we are concerned suggests one of three possibilities for the life history of the ancestor of vascular land plants. The algal model suggests that the ancestor was a simple gametophyte, with no independent sporophyte stage. The bryophyte example suggests a leafy megascopic gametophyte alternating with a dominant sporophyte

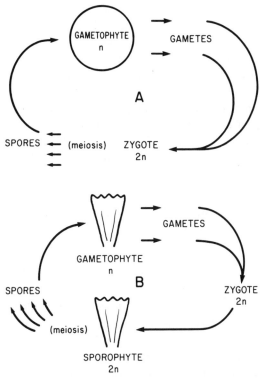

FIGURE 3.9 Schematic diagram illustrating zygotic life history (A) and sporic life history with isomorphic alternation of generations (B).

stage. Modern simple vascular plants suggest an ancestor that had already reduced the gametophyte stage to a microscopic, but nevertheless independent, form.

The three classes of earliest vascular plants—Rhyniopsida, Zosterophyllopsida, and Trimerophytopsida—all appear to have had the typical land plant life cycle with an independent macroscopic sporophyte. All of the recovered complete fossils are sporophytic, and no identifiable gametophytes of any size or morphology have been found. A possible exception is a putative gametophyte of *Rhynia gwynne-vaughanii* (Lemoigne 1968). If this interpretation is correct then *Rhynia*, and presumably others, had life histories similar to extant primitive land plants. While this observation accords nicely with the observed life histories of extant primitive land plants (e.g., Lycopsida, Sphenopsida, Psilopsida, Filicopsida), it creates difficulty on another front. In a later section we point out that simple green algae with a zygotic life history (figure 3.9) represent a likely ancestral form. These algae have no independent, macroscopic sporophyte generation. In the evolutionary sequence to the land plants, however, the nearest known "descendants," if they can be called such, are plants like *Cooksonia* or *Rhynia:* plants with an apparent sporophyte-dominant life history. Thus the problem of the evolution of the life history becomes a key issue in the evolution of the land flora.

3.5.2 The Question of Bryophytes and Fossil Gametophytes

Even if unambiguous gametophytes are recovered from the fossil record, it will probably be impossible to establish their relationship to a known sporophyte. Much of the interpretation will then rest on analogies to modern land plants and presumptive algal ancestors.

It is not difficult to explain and understand the absence of rhyniopsid, zosterophyllopsid, and trimerophytopsid gametophytes. It is reasonable to assume that they were microscopic, fleshy, subterranean thalli that would leave very poor fossil records, if any, and even then would be difficult to identify as true gametophytes.

The presumptive ancestral algae are represented today by taxa that have zygotic life histories. Because these ancestors presumably lacked separate sporophyte generations, the subsequent appearance and increasing dominance of the sporophytes demands a strong explanation. It is also essential to account for the forms of the algal gametophytes and their fates. Were they perpetually small microscopic forms or did evolution bring the advent of larger forms followed by their reduction again as the sporophyte became the dominant generation?

A few other presumptive gametophyte fossils are known: e.g., *Lyonophyton, Sciadophyton, Sporogonites.* They could be either those of bryophytes or the gametophyte stages of presumptive intermediate land plants. These gametophytes are early Devonian (Gensell and Andrews 1984; Thomas and Spicer 1987). *Sporogonites* (figure 3.10) probably is a fossil bryophyte. The sporangium, terminal on a stalk,

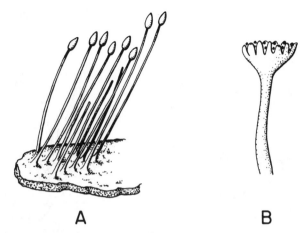

          A   B

FIGURE 3.10 Reconstruction of *Sporogonites* (A) and *Lyonophyton*
(B). *After Andrews 1960 (A) and Remy and Remy 1980a (B).*

arising in turn from what appears to be a prostrate thallus, is reminiscent of a hornwort such as *Anthoceros*.

Lyonophyton (figure 3.10) and *Sciadophyton* are problematical. The
former is regarded as an independent upright gametophyte (Remy
and Remy 1980a, 1980b). This gametophyte has no apparent connection with a dependent sporophyte. If this interpretation is correct,
then *Lyonophyton* cannot be a bryophyte. What then is the sporophyte? Remy and Remy (1980a) proposed *Horneophyton,* which is a
rhyniophytoid. There is no proof, but it is an interesting speculation
that has impact on our interpretation of the evolution of life histories
of land plants. *Sciadophyton* (Remy et al. 1980) also seems to be an
independent gametophyte. There have been no suggestions as to a
possible sporophyte stage for *Sciadophyton.* Remy (1982) subsequently
placed these putative gametophytes at the base of the evolutionary
dichotomy leading separately to vascular plants and bryophytes (i.e.,
among a hypothetical ancestral stock to these two groups). We
should not forget the third possible explanation for *Lyonophyton* and
Sciadophyton—namely that they evolved directly from some algal ancestry or very early land plants but are in fact evolutionary dead
ends.

Overall, the appearance of the terrestrial plant life history, sporophyte dominant, creates one of the major problems in elucidating the

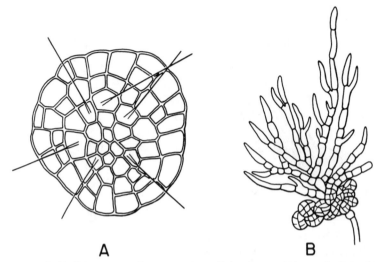

A   B

FIGURE 3.11 Diagram of vegetative *Coleochaete* (A) and *Fritschiella* (B). *(A) After Smith 1955. (B) After Singh 1941.*

sequence of events in the invasion. Whichever group among the algae is proposed as a likely type for the ancestral stock forces a number of assumptions that have not lent themselves to testing or verification.

3.6 Identifying the Algal Precursor of Vascular Plants

Initially, it would appear to be a relatively simple exercise to identify the putative precursor flora. First, look for living photosynthetic organisms that are simple and aquatic, and whose biochemistry resembles that of today's "primitive" (or simple) land plants. Next, establish that this identified group resembles land plants in key ultrastructural features; reinforce the assumptions with independent data (e.g., oligonucleotide and protein/enzyme gene sequences, cladistic analysis). Finally, establish from the fossil record and molecular dating that the presumptive ancestors existed prior to the postulated land invasion.

Assembling the data, subjecting them to detailed phylogenetic analyses, and at the same time eliminating other contenders should produce a consensus agreement on the presumptive ancestor. Indeed, these approaches have been highly successful in identifying

Charophyceae of the green algae (*sensu* Mattox and Stewart 1984), as the logical group in which to search for the putative ancestors. Although other workers clearly pointed the way (e.g., Pickett-Heaps and Marchant 1972), Stewart and Mattox (1975, 1978) examined the total sum of the data, with Mishler and Churchill (1985), Sluiman (1985), Bremer (1985), and Bremer et al. (1987) later providing the more detailed cladistic analyses.

Progress was slow and halting until the techniques of electron microscopy, modern comparative biochemistry and enzymology, protein and oligonucleotide sequencing, and biogeochemistry were applied to this evolutionary problem—all essentially within the last twenty years. The earliest phylogenetic attempts relied entirely on gross morphology. The ideas of Haeckel (1866) serve as an excellent illustration. Using gross morphology, Haeckel concluded that the Anthophyta, Pteridophyta, and Bryophyta together represented one major evolutionary branch, evolving from an ancestral group whose close evolutionary offspring included lichens, characean green algae, and red and fucoidean brown algae.

We now know with the benefit of hindsight that gross morphology is a treacherous and unreliable criterion. During the first third of this century basic biochemical knowledge completely altered these Haeckelian ideas. A similarity with land plants in chloroplast pigments (chlorophylls *a* and *b*, zeaxanthin, lutein, violaxanthin, neoxanthin), starch storage reserves, and cellulosic cell walls identified the green algae as the most likely progenitors. By corollary, the lack of biochemical similarity between land plants and other algal groups, e.g., the brown algae, quickly eliminated those groups as serious contenders. Lacking the benefits of modern ultrastructural, biochemical, and molecular analysis, students of the land invasion addressed the issues by looking at green algal morphology and life histories, and fitting these to the then current assumptions about the form and life history of the early land plants. The consideration of life histories was frequently the dominant criterion, and it became the guiding force in formulating ideas.

3.6.1 Early Ideas

In the absence of strong persuasive evidence to the contrary, we can argue that the primitive life history among the green algae is the zygotic form: i.e., the haploid gametophyte generation is primitive and meiosis occurs at zygote germination. These ideas have been influenced by the accepted and rarely challenged idea that simple gross morphology (a single cell) equates to evolutionary primitiveness. The correlation extends to life histories, as single-celled flagellates frequently have zygotic life histories.

All land plants, however, demonstrate alternation of generations, with the sporophyte as the dominant generation. (The bryophytes are the sole exception, as they demonstrate a sporophyte dependent on the gametophyte.) Within the green algae zygotic life histories occur typically in the Chlorophyceae and Charophyceae. Sporic life histories (figure 3.9) involving a gametophyte/sporophyte alternation of generations with meiosis at spore formation are most common in many of the Ulvophyceae. (A third gametic life history with meiosis at gamete formation—i.e., diploid gametophyte—is typical of the Caulerpales/Codiales assemblage within the Ulvophyceae. This life history is not considered here.) The origin of the sporophyte in land plants and its rise to dominance is thus a topic of great interest and importance.

The *antithetic* or *interpolation hypothesis* proposed by Bower (1908, 1935) held that the sporophyte generation was a new phase, interpolated between successive gametophytes, and was originally morphologically dissimilar to the gametophyte. This new phase arose from a delay in meiosis in the zygote, which previously had undergone a series of mitotic divisions to produce a simple sporophyte, probably dependent upon, and obviously heteromorphic to, the gametophyte. Bower developed his arguments based particularly on the need for a land plant to secure a method of reproduction that did not depend on repeated syngamy in a water medium (i.e., motile gametes). This would be achieved by airborne spores. Such a hypothesis implies that the early algal ancestors of the land plants possessed zygotic types of life histories.

At the time Bower developed these ideas the haplobiontic and diplobiontic life histories were known among green algae. Interestingly, Bower, in referring to *Coleochaete* and *Chara*, comments that "none of these seems to have hit on the innovation of postponing meiosis by interpolation of a diploid phase. . . . Here lies the biological gap between green aquatic and green amphibian life" (1935:489). However, it is important to realize that when this theory was developed the notion was widely accepted that the Archegoniatae (which includes the bryophytes) were the primitive land flora. This notion undoubtedly influenced the development of other ideas. It is easy to understand how the bryophytes, with their dependent sporophyte and semiterrestrial, nonvascularized form, were considered as obvious vascular plant ancestors. But it is now generally believed that the bryophytes do not occupy this position, and any current theory must take this into account.

An alternative theory, known as the *homologous* or *transformation theory*, proposed that the gametophyte and sporophyte generations of land plants arose from an ancestral green alga that showed isomorphic alternation of generations that were already independent. Such life histories do occur in extant members of the Ulvophyceae. Whereas this theory had been proposed in the late 1800s, Fritsch (1916) was the first to carefully analyze its merits. It is interesting to note that even then Fritsch had identified the Ulotrichales and Chaetophorales (*sensu* Fritsch), the latter which included *Coleochaete*, as the possible ancestors.

Both Fritsch (1945) and Zimmermann (1952) were proponents of the transformation theory, and they expanded these ideas into the realm of ancestral morphologies and early transmigratory morphological evolution. Fritsch, in particular, emphasized the importance of the heterotrichous thallus in the ancestral green algae, whereby erect branches develop from a prostrate portion. Fritsch again pinpointed the Chaetophorales as a likely ancestral stock and in particular a small green algae, *Fritschiella* (figure 3.10). *Fritschiella* had been discovered by Iyengar (1932) in India from damp pond mud, and it was shown by Singh (1941) to have an isomorphic alternation of generations. Fritsch observed, "that *Fritschiella*, in a somewhat specialized form, illustrates the existence of potentialities for a further elaboration in a Chaetophoralean type in the direction above postulated for an early land transmigrant" (1945:6). In identifying the Chaetophorales as it was then known, Fritsch was correct, but partly for the wrong reasons. More important, however, is that the hypothesis of the ancestors being from the Ulotrichales/Chaetophorales assemblage was not seriously challenged.

The next twenty-five years produced few changes or advances in ideas about the ancestral origin of the vascular land flora. Beginning in the 1960s the application of electron microscopy to studying events

both in cytokinesis and the structures of flagellar bases and roots had
a profound and far-reaching impact on green algal systematics and
views about the relationships of vascular land plants to putative
green algal ancestors.

3.6.2 Current Ideas

In the early 1970s Pickett-Heaps and colleagues undertook an extensive investigation of cytokinesis in green algae (reviewed by Pickett-
Heaps, 1975); they identified two fundamental forms (figure 3.12):

- *The phycoplast.* Cell division involves a cleavage furrow with a
  plate of microtubules in the plane of division, at right angles to
  the spindle microtubules.
- *The phragmoplast.* The microtubules are oriented at right angles to
  the plane of the new cell wall, which is formed from a plane of
  vesicles. This type is typical of embryophytes (vascular land
  plants) and such algae as *Chara, Coleochaete,* and *Klebsormidium.*

Pickett-Heaps and Marchant (1972) recognized the significance of this
difference and proposed a radically new classification of green algae.
They aligned such orders as the Charales, Zygnematales, some members of the Chaetophorales, and the genera in what later became the
Klebsormidiales into one evolutionary group leading to the higher
plants. This fundamental realignment of the green algae has received
support from work on flagella bases and roots (summaries in Melkonian 1982, 1984; Moestrup 1978; O'Kelly and Floyd 1984) and from
work in comparative biochemistry (summaries in Ragan and Chapman 1978; De Jesus et al. 1989).

Although there is still disagreement about green algal classification, particularly among the monadal forms (see Melkonian 1980;
Mattox and Stewart 1984; Steinman 1985; Van den Hoek et al. 1988;
Kantz et al. 1990) and among the Ulvophyceae (Zechman et al. 1990),
the classification of Mattox and Stewart (1984) represents a starting
point for developing a consensus classification.

We present our own preferred classification of plant groups (including algae) in table 3.1. We caution, however, that this classification is presented only as a framework for the purposes of this book.
The boundaries of Ulvophyceae, Micromonadophyceae, and Pleura-

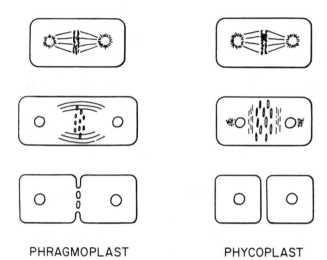

PHRAGMOPLAST   PHYCOPLAST

FIGURE 3.12 Diagrammatic illustration of phragmoplast (A) and phycoplast (B)during cell division. *After Lee 1989.*

strophyceae will certainly change in the future (see Kantz et al. 1990)
as more data become available. We are more confident that our
classification of the Charophyceae, with constituent orders and genera, will hold. This classification serves our purpose. It delineates
the "new" Charophyceae from which the vascular land flora almost
certainly arose. There are no strong arguments for such an ancestry
in any other class.

The biochemical data provide independent evidence to support
this view. Cladistic analyses of characters of green algae and vascular
land plants (Mishler and Churchill 1985; Bremer 1985; Bremer et al.
1985; Sluiman 1988; Theriot 1988) lend further support. A molecular
approach to the issue of land plant ancestry, using 5S rRNA nucleotide sequences, has been presented by Hori et al. (1985), Devereaux
et al. (1990), and van der Peer et al. (1990). While there are definite
limits to the usefulness of 5S rRNA sequences, owing to their small
size and highly conserved nature, the data support our conclusions.
More recent studies on small and large subunit rRNA sequences
provide better substantiation (Buchheim et al. 1990; Zechman et al.
1990; Krantz et al. 1990; Chapman and Buchheim 1992). A comparison focused on the analysis of cellular structures has been provided

by Graham and Kaneko (1991); that work clearly identifies the need for further reconsideration of green algal taxonomy, but it does not alter the status of the Charophyceae–vascular plant relationship.

It is important to emphasize that the charophycean ancestry hypothesis is by no means proven. This class has acquired the designation of the ancestral class partly by elimination of other groups and partly by positive correlation. Nevertheless, this conclusion is solidly based on biochemistry, molecular protein, gene nucleotide and rRNA sequence data, and cellular ultrastructure.

3.6.3 Potential Progenitors in the Charophyceae

How "close" are extant Charophyceae to our notions of the early migratory flora and land plants in terms of morphology and life history? Can we reasonably account for this charophycean–land plant relationship or should we look for, or hypothesize, another hitherto unknown group, or perhaps a totally different form of early land plant?

Unfortunately there is still a wide morphological and life history "gap" between extant Charophyceae, extant land flora, and presumptive primitive land flora. A number of green algae whose morphology and life histories are appropriate for a potential ancestor fail the biochemical and molecular tests of phylogenetic relationship. Most notable among these is *Fritschiella*, which had earlier acquired contender status, and the Ulotrichales. All those green algae now assigned to the other four classes are no longer credible candidates as land plant ancestors.

An examination of the five orders in the Charophyceae quickly limits our options. The sexual reproductive mechanisms of the Zygnematales and Charales make it difficult to argue a case for possible ancestral roles. The *Chara/Nitella* assemblage within the Charales is too complex and specialized morphologically and reproductively to represent an ancestral stock for the land plants. They have a macrophytic upright "stemlike" morphology in their gametophyte stage, but there is no evidence that the first land plants were erect plants. The pioneers were far more likely to have been prostrate leaflike structures. The antherozoid (male) of *Chara/Nitella* is flagellated, like that of ferns and other primitive land plants (e.g., fern allies), but it is morphologically dissimilar to them or algae, and sexual reproduction is oogamous. Moreover, the male and female structures of *Chara/Nitella* are much more specialized than those of the simple land plant gametophyte. The free zygote develops directly into a new gametophyte. It is thus difficult to imagine the life history and reproductive morphology of the Charales giving rise to a hypothetical simple land plant. Likewise the vegetative morphology does not fit in with any concepts of early land plant morphology based on living or fossil plants.

It is even more difficult to imagine representatives of the Zygnematales ("conjugating green algae") as potential progenitors. They do not possess flagellated reproductive cells. The life history involves fusion of vegetative cells, followed by internal fusion of amoeboid gametes. The zygote either germinates directly (after meiosis and disintegration of three nuclei), becoming a new gametophyte in the filamentous forms, or splits into four new daughter cells in the single-celled desmid families.

The Klebsormidiales are, from a life history perspective, weak candidates. They are very simple filaments with isogamous sexual reproduction and an apparent lack of an alternation of generations. Asexual reproduction of the haploid phase is known, but there appear to be no distinct structures for asexual reproduction. The Klebsormidiales might appeal to proponents of the homologous/transformation theory and to those (e.g., Schuster 1983a) who look for the algal ancestor to have had an isomorphic alternation of generations, but only if such alternation were to be demonstrated. However, the morphological and reproductive distance to the land plants is perhaps greater than that involving the Coleochaetales.

The Chlorokybales are poorly understood. It is a monotypic order whose single genus shows a sarcinoid growth form.

Overall—and more by elimination than by positive correlation—the mantle of closest extant algal relative to the putative early land plant has fallen upon the Coleochaetales. Is this a reasonable assumption? Could an alga resembling a modern Coleochaetales genus have given rise to a migratory form?

3.6.4 Coleochaetales as the Presumptive Progenitor

Within the order Coleochaetales, the genus *Coleochaete* appears the most likely model (figure 3.11). (Interestingly, in the earlier classifications this genus was put in the same order, Chaetophorales, as *Fritschiella*.) Schuster (1983a) has pointed out the pitfalls and problems of working with extant algae when trying to deduce the evolutionary development of bryophytes and vascular plants. Notably, the fossil record is poor, so the focus must turn to extant algae. One must use modern forms as analogues for building a likely scenario. We recognize the problems of using a modern *Coleochaete* as a model for the hypothetical ancestors. In the absence of alternative approaches or unrestrained imagination we would not decry it as "a waste of intellectual energy to try to derive land plants, and specifically hepatics, from *any extant group* of algae, except in the most general sense" (Schuster 1983a:895).

While we do not know the morphology of the earliest land plants we can assume they were much simpler than the upright forms represented by *Cooksonia*. Such plants were possibly foliose and attached to the substrate; they were probably cutinized and may even have had pores or simple stomatal structures, but they almost surely lacked vascularization. Nevertheless, we acknowledge that these assumptions may have been prejudiced by the early (now mostly discredited) notions of bryophytes representing typical early land plants and by our tendencies to fit evolutionary history to existing and known fossil morphologies.

A *Coleochaete*-type is plausible as a model of a potential land migrator. The thallus is a flattened cellular disc. Growth is by means of a peripheral meristem, and there is rudimentary parenchymous tissue (Graham 1982). It is not difficult to imagine the development of this alga into a form adapted for land—i.e., a simple "leafy" disc, with rhizoids for attachment to the soil. However, we must remember that this is a gametophyte generation. The gametophyte could have evolved into a form with pores and then a "leaf" that became cutinized and polystromatic. From the viewpoint of gross morphology this would imply an evolution to a bryophyte-like form. But, as will be shown later in this chapter, the bryophytes are not part of the evolutionary lineage to the land plants. In the most primitive of extant vascular land plants the gametophyte is very small and subterranean. It is not difficult, therefore, to postulate the evolution of *Coleochaete* into the gametophyte stage of a primitive land plant, especially if we use the gametophyte of modern lycopods as the example.

Sexual reproduction in *Coleochaete* is oogamous, meaning fertilization of egg by a sperm. Eggs and flagellated sperm are produced at the edge and fertilization occurs on the female gametophyte. The diploid zygote is retained on the haploid gametophyte. In this respect *Coleochaete*, where the diploid zygote is also retained within the gametophytic tissue, is exceptional among green algae (Graham 1985). Graham has discussed the necessity of this scenario vis-á-vis the land plant reproductive system. Simple, primitive land plants show oogamous reproduction, with retention of the zygote. There is a nutritional relationship between the developing zygote and its gametophytic retainer.

Evolutionary invasion of dry terrestrial environments, via the intermediacy of damp milieu, necessitates the eventual elimination of reproductive systems dependent on water. The gradual disappearance from the life history of flagellated gametes and zoospores is thus a necessary postulate. Certainly reproduction with flagellated cells could still exist in terrestrial plants, provided the gametophytes resided as subsurface or surface forms capable of producing and liberating sperm during wet periods when surface films of water were present. The evolution of a *Coleochaete*-form to a surface/subsurface gametophyte is plausible.

In dry terrestrial habitats, sexual reproduction that involves the transfer or a movement of one reproductive cell (male) to another (female) on separate plants is difficult. Asexual reproduction can be less dependent on the milieu. This potential for reduced fertilization rates or their likelihood prompted Searles (1980) to suggest that the development of the sporophyte in the terrestrial habitat was driven by a need to compensate for the drawbacks of sexual reproduction in terrestrial systems. Production of large numbers of spores derived whenever a single fertilization managed to succeed would help achieve this. Water-independent dissemination (i.e., wind-borne or ballistic discharge) of spores would be a further advantage.

Thus we have a possible scenario for the origin of the terrestrial gametophyte. We must now provide for the origin of the sporophyte from a *Coleochaete*-like life history.

Coleochaete does not show an alternation of generations like that of land plants, but this does not represent a major theoretical hurdle. The zygote undergoes division meiotically to produce flagellated zoospores that develop directly into new gametophytes. A necessary preadaptation for the life history of the land plants would require further development of the zygote into a small diploid and hetero-morphic sporophyte that did not immediately divide into zoospores and was initially dependent on the gametophyte. Subsequent acqui-sition of independence for the mature sporophyte and its develop-ment into the dominant generation with water-independent repro-duction (i.e., walled spores) would complete the scenario for the land plant life history.

This scenario is, in fact, the *antithetic* or *interpolation theory* for the origin of the land plant life cycle. Although this theory currently is in vogue, it is important to remember that compelling evidence against the homologous or transformation theory has yet to be produced. Schuster (1983) argued for a common ancestor to bryophytes and vascular plants that possessed an isomorphic alternation of genera-tions. But the failure to find this life cycle in any of the extant Charophyceae has probably played a role in marshaling support for the antithetic hypothesis.

The proposed sequence of the antithetic hypothesis does not re-quire the *immediate* development of a system of sexual reproduction that is independent of a water medium. The early land flora presum-ably were semiaquatic, and thus were not pressured for the immedi-ate elimination of water-mediated mechanisms of sexual reproduc-tion. This argument is supported by the fact that a reproductive style in which flagellated sperm depend on a film of water is found in many of today's primitive land plant groups: Lycopsida, Sphenop-sida, and Filicopsida. An alga with a life cycle that had the *potential* to evolve into a type at least partially suited to a terrestrial or quasi-terrestrial environment would have had a preadaptive advantage. Here, too, *Coleochaete* fits the bill. In the absence of an as-yet-undiscovered green alga, or a progenitor "designed" to meet our own preconceptions, the extant *Coleochaete* is indeed the best model to explain some of changes that had to occur during the land mi-gration.

We cannot emphasize enough that the difficulty of fitting the homologous or transformation theory into a hypothesis for the origin of the land flora should not induce one to disregard that hypothe-sis—or to fail to look for alternatives. After all, very few extant organisms are even remotely attractive as models; there are problems in extrapolating back 450 million years; and the fossil record is scant and demonstrates a wide gap between vascular land plants and whatever we postulate as the algal ancestor. Rather than trying to prove one particular hypothesis, it is perhaps most prudent to marshal evidence to disprove others.

3.6.5 The *Coleochaete* Type as the Ancestor: Some Caveats

The scenarios we have presented for land invasions and the deductions we have made about the nature of the green algal ancestor are all based on our ideas of the presumptive characteristics of the probable early land plant and our tendency to fit our hypotheses to known extant and fossil forms. We have anointed a *Coleochaete*-like alga as the possible ancestral type largely by default. We simply do not have a better, more viable candidate that meets the morphological and life history concerns.

In making this decision we are in effect suggesting that a *Coleochaete*-like alga existed at least as far back as the onset of land colonization. The possibility of convergence has not been factored into the hypotheses. At this time there are very few data bearing on the possibility of convergence. Phylogenetic trees based on oligonucleotide (small subunit rRNA) sequences should help resolve the issue. The best information that we currently have, based on SSU rRNA sequences (Zechman et al. 1990; Krantz et al. 1990; Chapman and Buchheim 1992), shows the following: Higher plants and *Equisetum* (Sphenopsida: horsetails) share a common evolutionary ancestor with *Klebsormidium*, but also with *Chara* and *Nitella*. Recent partial LSU and SSU rRNA sequences for *Coleochaete* indicate that this genus is *less* closely related to higher plants than are the Charales or Klebsormidiales (Chapman and Buchheim 1992). Bremer (1985), Mishler and Churchill (1985), and Bremer et al. (1987) have placed the Coleochaetales, rather than the Charales, as the sister group to the land plants. The Klebsormidiales are further removed. However, these authors caution that more species of *Coleochaete* need examination and that homoplasy occurs in both the Charales and Coleochaetales.

Another caveat is ecophysiological. The land invasion of plants, as currently held, involved a freshwater stage intermediate between the marine and the fully terrestrial. Raven (1985) has pointed out the problems with such a transition if one examines the physiology of extant plants. All known charophyceans are freshwater, as are most Chlorophyceae. The great majority of marine green algae belong to the Ulvophyceae and Micromonadophyceae, which are only very distantly related to the Charophyceae/land plant complex (Zechman et al. 1990; Krantz et al. 1990). There have been claims (Racki 1982) for a marine habit for the earliest members of the Charales, which have been found in Devonian rocks. However, the significance is questionable. These fossils are true *Chara/Nitella* types, clearly not on the path to a land flora.

Adherence to the Coleochaetales-type or Klebsormidiales origin of the land plants suggests that a marine *Coleochaete*-type ancestor was entrapped in a brackish and then eventually aquatic nonmarine situation. Perhaps these environments began as ponds shoreward of high-tide impoundments and that became more and more fresh as sea levels dropped. Marine charophyceans are unknown, and there is no compelling evidence that the early land plants of the fossil record inhabited a marine environment.

We have used modern phylogenetic methods of data acquisition and interpretation to identify a possible ancestral group that can be related to modern relatives. However, different data sets and analyses produce somewhat conflicting results. It is clear that in the aggregate none of the three candidate groups provides a totally comfortable solution. This is perhaps to be expected. There is nothing to negate the possibility that a long-disappeared charalean, klebsormidealean, or coleochaetalean is closer to the land plants. This could also apply to the possibility of a long extinct group closely related to these three contending orders. Despite these caveats *it appears that our best models among extant algae come from either the Klebsormidiales or Coleochaetales*. We now ask: Are there alternative hypotheses, and does the fossil record support any particular hypothesis?

3.7.6 Alternative Hypotheses

If the idea is accepted that a Coleochaetales- or Klebsormidiales-like alga represents a possible ancestral stock, then it follows that the migration involved a multicellular alga. Stebbins and Hill (1980) have argued an alternative scenario, namely that the land invasion was carried out by single-celled algae and that the current (freshwater) charophyceans are secondarily aquatic. This hypothesis does not alter the belief that the ancestral group was either the Charophyceae or some hitherto unknown group that left no modern descendants or whose descendants have not yet been discovered. In promulgating this hypothesis Stebbins and Hill proposed that the ancestor was diplobiontic and isomorphic; this hypothesis therefore invokes the homologous theory. They believe that these unicellular forms evolved into multicellular forms living on damp substrate surfaces.

Graham (1984, 1985) has considered this argument and has pointed out the problems, at least from the life history aspect. Fundamentally, there are no charophycean algae with both oogamous sexual reproduction and alternation of generations. Alternation of generations in algae requires that the gametophyte and sporophyte be independent. Accordingly, the Stebbins and Hill hypothesis would posit an acquired retention/dependence of the zygote, and a gametophyte that presumably became reduced in size or stayed microscopic. But this is the reverse of what is generally believed to have happened. Graham has summarized the controversy, "It is difficult to conceive of any selective advantage for the multistep evolutionary pathway required to generate a land-plant life cycle from a diplobiontic one . . . collectively observations support the hypothesis that land plants arose from advanced charophycean algae by a single evolutionary step, a delay in meiosis" (1985:183). The Stebbins and Hill hypothesis would also require a further shifting of the phylogenetic positions of the bryophytes.

Despite these problems, the Stebbins and Hill hypothesis cannot be ruled out. At least from the viewpoints of physiology, biochemistry, and morphology, the hypothesis that the land invasion was carried out by single-celled algae and that the current (freshwater) charophyceans are only secondarily aquatic is indeed plausible (see Raven 1985). It should be noted that there are few single-celled Charophyceae. *Stichococcus* in the Klebsormidiales is one such example. Members of this extant genus are all single cells or filaments readily dissociable into single cells. It is a soil, freshwater, or estuarine genus lacking sexual or zoosporic reproduction. This is a far cry from what we imagine the earliest land plants were: a simple prostrate morphology, with a small sporophyte developing on the gametophyte, followed by evolutionary change to the system currently existing in primitive land plants, i.e., a small subterranean gametophyte on which the eventually much larger sporophyte initially develops.

The dispute may be impossible to resolve until researchers are willing to seriously entertain the possibility of multiple attempts at migration involving different ancestors—with multiple failures and even multiple successes occurring over a long time span—and to admit the possibility of ancestors whose descendants have gone extinct or remain undiscovered. We have formulated a working scenario that does take account of these likely complications.

3.7 The Fossil Record of Green Algae

We now ask the question, What does the fossil record tell us, and is it in agreement with the proposed sequence of events? Unfortunately the record is hazy and difficult to interpret.

One could expect a gradual continuum beginning with the presumed ancestral algae to the earliest semiaquatic plants and finally to true terrestrial plants. If typical features of terrestrial adaptation—i.e., tracheids or conducting strands, trilete spores and tetrads, and stomata—are absent in a given fossil, it is very difficult to discern conclusively whether that fossil is an advanced aquatic algal genus or a very simple land plant. One must remember that many simple algae, including Coleochaetales, do not have submersion as a growth and survival *requirement* but are able to live and propagate quite readily in air, as on damp agar. The sedimentary record does not allow identification of the habitat of early plants with absolute certainty. Fluvial outflow may have washed plant remains into marine sediments, and the confluence of freshwater and marine environments would be difficult to identify.

Furthermore one must not fall into the trap of attempting to fit every fossil genus into an evolutionary continuum. There were almost certainly evolutionary failures: algae/plants that gained a precarious foothold in the initial semiaquatic environment but then became extinct with no descendants. We will, in fact, show that some fossils are best explained as outright evolutionary failures.

3.7.1 The Time Frame

Despite the fact that it is probably impossible to unequivocally identify charophycean algae in the fossil record, one can find indirect evidence for the presence of "green algae" in the Cambrian and

Ordovician—and even the late Proterozoic. Presumptive calcified green algae of the Ulvophyceae, similar to extant genera of the Codiales and Dasycladales, have been identified from the late Precambrian to Ordovician. Examples include *Cyclocrinites*, *Vermiporella*, *Archaeobatophora*, *Rhabdoporella*, and *Primicorallina* (Herak et al. 1975; Elliott 1989; Johnson and Sheehan 1983). Although the Ulvophyceae themselves are not ancestral to the higher plants, this fossil record indicates that Chlorophyta were well established by the very late Proterozoic or early Cambrian. Fossil oogonia of the Charales-Charophyta have been identified in the early Devonian to late Silurian, suggesting that Charophyceae were established by the Silurian. Edwards and Lyon (1983) have recovered a putative Charophyceae (*Palaeonitella*) and Ulvophyceae (*Mackiella* of the Ulotrichales) from the early Devonian. This fossil record by itself does not demonstrate that charophycean algae existed in the Cambrian or Ordovician, or even earlier.

Sequences of SSU, LSU, and 5S rRNA may permit a measure of quantification, as genetic distances on phylogenetic trees. The most complete and reliable analysis of the green algae based on SSU rRNA sequences demonstrates that the line leading to the Charophyceae and land plants diverged early from the evolutionary lines leading to the Ulvophyceae (Zechman et al. 1990; Chapman and Bucheim 1992). The fossil record of the Dasycladales/Caulerpales (Ulvophyceae) extends at least as far back as the Ordovician. The green algal lineage may be as old as 800 million years (A. Knoll, quoted in Chapman and Buchheim 1992). It is reasonable to argue that members of Charophyceae existed prior to our hypothesized time (about 450 mya) for the emergence of a land flora.

One new approach is the application of molecular biological techniques for determining evolutionary distances and then applying a time scale. This is the molecular clock hypothesis proposed first for proteins and later for nucleotides. For a given protein or nucleotide, it is argued that the rate of substitutions is constant in all lineages. The hypothesis is highly controversial (see Li and Graur 1990; Wilson et al. 1989). The principal stumbling blocks are the assumption about rate constancy, the problem of widely differing generation times, and whether or not the principle can be applied only to closely related groups or to all groups regardless of the extent of evolutionary separation. Despite the controversy, it is interesting and possibly instruc-

tive to apply the molecular clock to green plants (green algae, bryophytes, and vascular plants) that must trace back to a common
origin.

Ramshaw et al. (1974) used cytochrome c sequence data to place
the divergence of the land plants from the Ulvophyceae at 700 mya.
Calculations with the nucleotide sequence of the cytochrome c gene
(Amati et al. 1988) provided a date of 700–750 mya for the separation
of the land plants from *Chlamydomonas* (Chlorophyceae). An extrapo-
lation involving 5S rRNA sequences of Hori and Osawa (1987) places
the divergence of the combined Charophyceae and land plant lineages from other green algae at 800 mya; the Charophyceae and land
plants then diverged from each other at 550 Mya. The move of
the Charophyceae to the land presumably occurred after this class
diverged from the other green plant groups.

The fossil record indicates that a land flora, or at least the invaders, were establishing themselves probably no later than 450 mya
years ago. Molecular evidence—controversy about interpretations
notwithstanding—suggests that green plants may have diverged into
their principal groups about 200–250 million years earlier, or about
700 mya years ago. And there is a possible fossil record for green
algae extending back to even 800 mya.

3.7.2 A Survey of Fossil Genera

In the evolutionary sequence from the algal form to the first recognizable land plant there is a wide morphological chasm. Fossils that
cannot be unambiguously identified as either land plant or algae fall
into this void and have on occasion been referred to as algal-like.
Such forms are not necessarily representative of the green plant
lineage, either green algal or early land plant, and some might be
evolutionary dead ends.

Calcified fossil algae are much easier to identify or assign to a
major taxon than are noncalcified forms. Thus paleobotanists have
no real problem with many putative Ulvophyceae. But fleshy foliose
forms lacking land plant features can be enigmatic. They could be
green algae of a type logically evolving from the presumptive ancestral type. They could likewise be brown algae, reminiscent of foliose
Fucales, Dictyotales, or Scytosiphonales. This matter cannot be re-

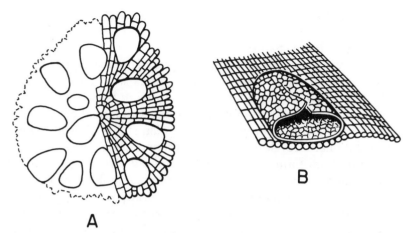

FIGURE 3.13 Reconstruction of *Parka decipiens* showing upper surface of whole plant (A) and section of plant showing putative sporangia (B). *After Taylor 1988.*

solved in the absence of reproductive structures. Then, too, enigmatic fossils could also be representatives of long-extinct groups, either aquatic or terrestrial, unrelated to any extant taxa. The interpretation is further complicated by the absence of any really complete fossils for these questionable specimens. Despite this gloomy assessment there are, however, a number of fossil genera from the Silurian and Devonian that can be classified with some precision: *Parka, Pachytheca, Protosalvinia, Spongiophyton, Palaeonitella, Prototaxites,* and *Nematothallus.*

Parka (figure 3.13) is known from the late Silurian and early Devonian as a small ovoid thallus, probably a few cells thick with a dorsoventral orientation, with numerous surface or embedded sporangia. Morphologically it can be considered an alga (Niklas 1976a). Considering only its gross morphology it is not difficult to imagine how *Parka* might have evolved from a Coleochaetales-Klebsormidiales ancestor, but an ancestry with some simple Ulvophyceae or brown algal relationship cannot be ruled out. The presence of sporangia, if that is what they are, implies a separate sporophyte generation. Perhaps *Parka* represents a failed experiment of a *Coleochaete*-like alga to evolve from its present form and life history to an isomorphic alternation of generations. Overall, the most likely interpretation for *Parka* on the basis of its morphology is that of a green alga (but with no proof that it is a charophycean).

Pachytheca (Niklas 1975) is more complex. The small spherical thallus of radial filaments shows both a cortex and a medulla. Like *Parka* it is known from the late Silurian through the early Devonian. However, *Pachytheca* does not provide much information about the evolutionary events leading to the land plants. It is difficult to imagine evolutionary development of the three-dimensional structure of this genus from forms represented in the putative ancestral groups of land plants or indeed from any charophycean unless one is willing to admit the possibility of a filamentous form producing something resembling *Cladophora* balls. But it is perhaps more plausible to imagine a relationship of *Pachytheca* to simple cushion forms of the Chlorophyceae and Ulvophyceae—perhaps a *Microdictyon*, or even a cushionlike *Codium*.

Protosalvinia (Phillips et al. 1972; Niklas and Phillips 1976; Niklas et al. 1976) is probably not a green alga, at least as we know them today. Tetrahedral spores of the type found in *Protosalvinia* are unknown from the green algae, and also from the brown algae to which this problematic genus has sometimes been assigned (Schopf 1976). Gray and Boucot (1979) have proposed that *Protosalvinia* is in fact a land plant. This view has support in the work by Romankiw et al. (1988), who identified phenolics, alkyl phenolics, and alkylbenzenes from *Protosalvinia* remains. The chemical signatures are very similar to those from coalified wood. Niklas and Chaloner (1976) had previously found presumptive lignin residues. The morphology, however, prevents any satisfactory pigeonholing. Overall, *Protosalvinia* is an organism that invites speculation (Stewart 1983), including the possibility that it may well represent another evolutionary experiment that failed and left no descendants.

Like *Protosalvinia*, *Spongiophyton* (figure 3.14) also has a history of being bounced from one taxon to another (Niklas and Chaloner 1976). The morphology is that of a cuticular envelope, of a few centimeters linear dimensions. The interesting feature, which would exclude any algal affinity, is the presence of pores on only one side— a side with a thick cuticle. The other side, lacking pores, has a thinner cuticle. The gross morphology is suggestive of a leafy thallus,

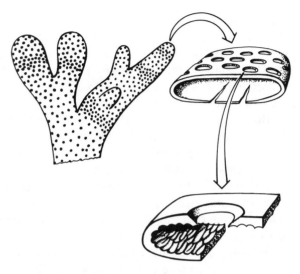

FIGURE   3.14 Reconstruction of *Spongiophyton nanun*. *After Taylor 1988.*

with the pores being an early adaptation for air exchange. Whether it is bryophytic, an "early land plant," or even algal in affinities must await the discovery of at least reproductive structures.

The earliest unequivocal charophycean is *Palaeonitella* (Edwards and Lyon 1983). This genus has been recovered from the Rhynie Chert (lower Devonian, Siegenian). While it does indicate that Charophytes were well developed by about 400 mya, it is of little help in bridging the gap between the presumptive ancestors and the earliest land plants, for which we need to go back in time at least another fifty million years.

Prototaxites and *Nematothallus* are two fossil puzzles that we mention only because they have at times been assigned an algal identity and because they are now best regarded as early evolutionary experiments in land colonization that failed. *Prototaxites* is a large organism (up to a meter wide and two meters long), with a parenchymatous structure composed of hyphae and tubes (Schmid 1976) and with septal pores in the filament walls. The genus has been assigned at some time to nearly every major primitive plant group. Each attempt has been based on stretches of the imagination and perhaps reflects a prejudice that every fossil must be allied to an extant group.

Nematothallus is the type genus of the order Nematophytales (Lang 1937) and, like *Prototaxites*, has a checkered taxonomic history, including suggested relationships to such green algae as *Codium* (see Strother 1988). The fossils themselves are fragmentary, composed of cuticle with associated tubes (Edwards and Rose 1984; Strother 1988). The appearance of cuticular pores suggests an adaptation for gas exchange and thus terrestrialization. There is no easily discernible relationship to any extant taxon (discovery of reproductive structures might help), so this order probably represents another example in the growing list of apparent evolutionary dead ends.

In summary, therefore, the fossil record does not reinforce the identification of any one particular algal group as the land plant ancestor. It does indicate that morphological and reproductive innovation was occurring in perhaps many lineages. It also suggests that there were many failed evolutionary experiments; researchers must therefore be more willing to take this into account. Then too, when considering the various land invasion hypotheses, one must not dismiss the possibility that the ancestry of vascular land plants lies in an as-yet unknown and undiscovered group.

3.7.3 The Paleobiochemical Record

Overall, the fossil record provided little help in identifying the sequence of events, or their timing, from the aquatic green algae to established land plants. *Parka* and *Pachytheca* indicate what could have happened, but there is no proof that they were in fact charophycean algae. Are there other data, besides the fossil record, that bear on the question of ancestry?

Some additional help can, in fact, be found in the paleobiochemical record. Earlier in this chapter we identified the functional adaptations needed for a land existence—the "four footprints" we called them. These adaptations carry with them the need for specific biochemistries. Aquatic green algae could thus be expected to have a different chemical signature from that of a primitive land plant. If the key compounds that distinguished the earliest land plants were not altered by fossilization, and if our knowledge of the comparative biochemistry of extant plants is good, it should be possible to put this biochemical data to use. Small amounts of original compounds, however, might disappear from the fossil record, giving a false indication of absence. And there is always the assumption that biochemistry of a group has remained essentially unchanged over time.

Phenols from the diagenesis and degradation of lignins would be a characteristic signature of vascular plants (except for bryophytes synthesizing dilignols). Cutin, present in higher plants and bryophytes, should leave a characteristic signature of C_{16} carboxylic acid. Certain terpenes, e.g., pentacyclic ursanes and lupanes, are characteristic of higher plants. In addition to looking for specific signatures, one can take an entire signature profile. In this case the assumption is that the aggregate of minor differences will produce signatures that "cluster" according to taxa. Using this approach, Niklas has begun to unravel the questionable relationships between the algae and the nonvascularized plant fossils (Niklas and Chaloner 1976; Niklas 1976a, 1976b, 1976c, 1979, 1980, 1981). However, the recent presentation of evidence for ligninlike compounds in reproductive *Coleochaete* (Delwiche et al. 1989) suggests that paleobiochemistry may be of only partial help and that much more detailed analysis is needed.

An examination of *Parka* revealed signature compounds more typical of green algae—ergostane, carboxylic acids predominating with C_{17} (ex hydrocarbon), and the predominance of even chained C_{12}, C_{16} acids. A similar examination of a supposed early land plant *Orestovia* (see also Romankiw et al. 1988) and to *Protosalvinia* and *Spongiophyton*, coupled with a comparison to the extant green alga *Botryococcus*, suggested an algal-like profile rather than that of a higher plant. *Solenites*, a higher plant megaspore, and *Taeniocrada*, a vascular plant, yielded hydroxy fatty acid signatures, which would be typical of higher plant cutin and suberin. However, all of the presumptive algal-like genera and *Orestovia* showed the presence of phenolics indicative of higher plants. *Spongiophyton*, which has some algal-like features, closely relates to *Orestovia*, *Nematothallus*, and *Prototaxites*. These four are nonvascular plants. The presence of hydroxy-fatty acid in *Prototaxites* suggests cutin, and thus terrestrialization.

As might be expected, *Rhynia*, *Cooksonia*, and *Renalia* formed a chemical cluster. Other higher plants not regarded as Rhyniophytes—e.g., *Gosslingia*, *Sawdonia*, *Zosterophyllum* and *Crenaticaulis*—are well separated by cluster analysis from the Rhyniophytes. This cluster pattern is shown in figure 3.15. All the "algae" and nonvascu-

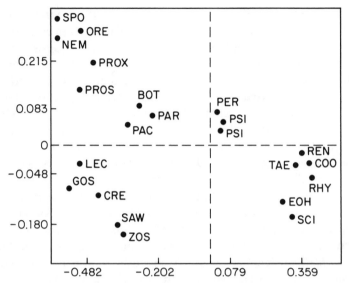

FIGURE 3.15 Coordinate diagram based on weighted chemical profiles of early plants. SPO = Sporogonites. ORE = Orestovia. NEM = Nematothallus. PROX = Prototaxites. PROS = Protosalvinia. BOT = Botryococcus. PAR = Parka. PAC = Pachytheca. LEC = Lecleria. GOS = Gosslingia. CRE = Crenaticaulis. SAW = Sawdonia. ZOS = Zosterophyllum. PER = Pertica. PSI = Psilophyton. REN = Renalia. TAE = Taeniocradia. COO = Cooksonia. RHY = Rhynia. EOH = Eohostimella. SCI = Sciadophyton. *After Niklas 1980.*

lar plants cluster in one quadrant. The only two genera that can legitimately be regarded as "algal" are *Parka* and *Pachytheca*, but there is still no compelling evidence to establish them as antecedents to higher plants. In combination with the morphology, these data lead us to conclude that there were many migrations occurring during the transition and that most eventually failed.

3.8 Phylogeny of Vascular Plants, Bryophytes, and Fungi

The preceding sections have described our proposed sequence of events in the pioneering of land by plants, along with the alternatives. Our sequence is based on assumptions for the necessary physiological and morphological adaptations needed for a land migration.

The ideas are tied to modern land plants and algae and accords well with the fossil record. If our proposed sequence is valid, then we should be able to examine the evolutionary relationships of the plants involved and suggest a phylogenetic scheme.

3.8.1 Early Vascular Plants

Current ideas group the early vascular land flora into three classes (Rhyniopsida, Trimerophytopsida, and Zosterophyllopsida) and the "rhyniophytoids," all of which are extinct. What are the relationships of these three classes to each other and to later classes of extant land plants? Before we offer our conclusions, a few comments on transitional forms are in order.

In any taxonomic scheme and phylogenetic continuum there will be plants that can be regarded as transitional forms between taxa. While transitional forms that display features of more than one of these three earliest classes may be a systematic nuisance, they are extremely valuable aids in developing phylogenies and tracing evolutionary development. The fossil record of early land plants reveals a number of such transitional, and therefore problematic, forms.

Renalia is one example. With terminal sporangia it resembles a rhyniopsid, but it also resembles a zosterophyllopsid because of the shape of the sporangia and their location along the distal edge. *Renalia* is, however, placed in the Rhyniopsida (Edwards and Edwards 1986). *Kaulangiophyton*, on the other hand, shows attributes of both Zosterophyllopsida (nonvascularized enations and terminal sporangia) and the lycopods (sporangial shape, mode of dehiscence, and exarchic protostele). The morphological evidence from the fossil record, which is all there is to work with, suggests that the class Zosterophyllopsida provided the transitional stock between some earliest land plants and the class Lycopsida (with both fossil and extant members). This is the view expressed by Taylor (1988), and it resembles those put forward by Chaloner and Sherrin (1979), Stewart (1983), and Gensell and Andrews (1987). (Figures 3.16 and 3.17 are visual presentations of this interpretation.) Taylor also placed the Trimerophytopsida as the transitional bridge to the Sphenopsida (which includes ferns and seed plants). *The trimerophytes thus occupy a central position for subsequent vascular plant evolution. Rhyniopsida represent an evolutionary dead end.*

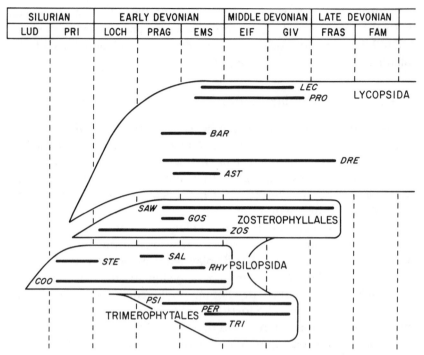

FIGURE 3.16 Postulated phylogenies of early land plants. LEC = Leclergia. PRO = Protolepidodendron. BAR = Baragwanathia. DRE = Drepanophycus. AST = Asteroxylon. SAW = Sawdonia. GOS = Gosslingia. ZOS = Zosterophyllum. SAL = Salopella. STE = Steganotheca. RHY = Rhynia. COO = Cooksonia. PSI = Psilophyton. PER = Pertica. TRI = Trimerophyton. *After Selden and Edwards 1989.*

Taylor (1988) has also proposed that all three classes—Zosterophyllopsida, Trimerophytopsida, and Rhyniopsida—evolved separately from a common ancestral stock that might have been similar to *Cooksonia*. Stewart (1983), however, has argued emphatically that the zosterophyllopsids represent the bridge between mid-Silurian Rhyniopsida and lower Devonian lycopods (e.g., *Asteroxylon, Drepanophycus, Baragwanathia*).

The fossil genus *Baragwanathia* illustrates some of the problems that can be encountered (Knoll et al. 1986). *Baragwanathia* is a vascular plant with clear affinities to the lycopods. Because of its morphologi-

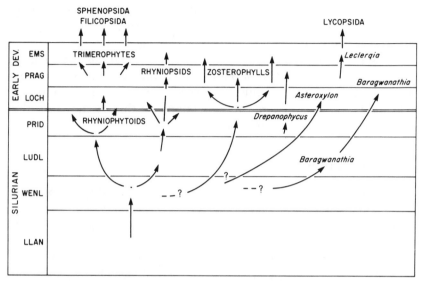

FIGURE 3.17 Postulated phylogenies of early land plants (cont.). *After Chaloner and Sheerin 1979.*

cal complexity it is presumably more advanced than other zostero-phyllopsids—if, indeed, it was a zosterophyllopsid. Specimens were originally described from Australia (Lang and Cookson 1935) and assigned a late Silurian age. Edwards et al. (1979) and Hueber (1983) questioned the age and argued for the early Devonian (390 mya). Garratt and coworkers continue to support a late Silurian age (Garratt 1978; Garratt and Rickards 1984; Garratt et al. 1984).

There is an interesting biogeographical component to the issue of classifying *Baragwanathia. Baragwanathia* is represented in both north-ern and southern hemispheres. The first appearance in the northern hemisphere is recorded from the early Devonian (Emsian). But con-sider that *Cooksonia* is unknown from the southern hemisphere. There is thus a situation where a plant that is morphologically more complex than *Cooksonia* appears essentially coeval with *Cooksonia*, but in a different hemisphere. This fact raises doubts about the possible origin of zosterophyllopsids from rhyniophytes and "cooksonioids." And this, in turn, should lead one to question whether vascular plants are indeed monophyletic (see Stewart 1983) and to admit the possibility that much earlier vascular plants are still to be found. There is, of course, the inevitable and recurring caveat: all these arguments hang on a slender thread of a sparse fossil record.

It should be noted, however, that Rayner (1988) has commented that *Dutoitea*, described from South Africa and currently considered a rhyniophytoid, is very similar to *Cooksonia*. He has raised the possibility that the two genera are congeneric. If that is the case, the biogeographical hurdle would no longer exist.

3.8.2 Bryophytes

Are there any extant plant groups whose phylogenetic position suggests a possible intermediacy between the green algae and the early vascular land flora? Specifically, are the bryophytes a possible intermediate group?

The revolution in green algal systematics demands a reconsideration of the phylogenetic status of the bryophytes. For at least the first half of this century bryophytes were considered as intermediate between green algae and vascular land plants. Their morphology, at least superficially, fitted into then-current views. Bryophytes inhabit damp environments. The gametophyte is leafy and often larger than the sporophyte. Both gametophytes and sporophytes have compound pores or simple stomata facilitating gas exchange. They lack water-conducting xylem, but mosses do have cells resembling sieve tubes and tracheids—which are referred to as leptoids and hydroids, respectively. Notably, the sporophyte generation is retained on the gametophyte, where it produces thick-walled spores totally independent of water for dissemination. The discovery of flavonoids and lignin-like dilignols or phenylpropanoid dimers (but not true lignin) tended to reinforce these assumptions of intermediacy for many years.

The bryophytes do show an alternation of generations with morphology presumptively preadapted for development to a typical terrestrial plant. However, the gametophyte is the longer-lived generation and the one that is morphologically more complex.

The suggestion of an evolution of land plants from charophycean algal forms via the bryophytes requires a scenario whereby the gametophyte evolved morphologically (in the bryophytes) and then became reduced to a subterranean simple thallus in the fern allies and ferns. This hypothesis of bryophytic intermediaries held sway for many years, not so much because of compelling evidence in its support, but rather because of a lack of strong contrary arguments. In the light of modern evidence based on the ultrastructure of reproductive morphogenesis and mitosis, combined with classical morphology, it now appears that the bryophytes (mosses, hepatics, and hornworts) are polyphyletic (Duckett 1986; Duckett and Renzaglia 1988). This fact by itself now makes it difficult to visualize evolutionary intermediacy for the bryophytes. There is also the possibility that they evolved from some ancestral stock with vascular land plants. (Crandall-Stotler 1980, 1986; Mishler and Churchill 1984, 1985; Duckett 1986; Smith 1986). Schuster (1983b) has in fact argued that the Anthocerotae (hornworts) may have derived from a distinct algal group that migrated to the land only in the late Devonian.

3.8.3 Fungi

The growth and maintenance of fungal mycelia and populations requires water (dampness or submergence) and a supply of organic carbon. Fungi play a vital role in the recycling of organic matter. Their appearance in the terrestrial ecosystem must have followed or paralleled the migration of plants and animals onto land. Fungi are known from the marine environment, but the great majority are fully terrestrial or inhabit freshwater. The phylogenetic relationships of the various fungal groups and the paleoecological aspects of their appearance on land (see Stubblefield and Taylor 1988; Sherwood-Pike and Gray 1985) will not be considered.

The interesting question is the role of the symbiotic mycorrhizal fungi in establishing a land flora. These fungi form dense sheaths that surround the roots. They contribute hormones to the plant and aid in mineral nutrition during periods of deficiency. The fungi benefit, in turn, by acquiring organic carbon from the roots. The advantage for healthy plant growth in the presence of mycorrhizae is well established.

The establishment of mycorrhizal interactions with rooted vascular plants would have conferred an advantage for early land plants. But the very first land plants may have existed without the benefit of mycorrhizae. Pirozynski and Malloch (1975) suggested that the massive colonization of land by plants in the Silurian and Devonian was possible because of this mycorrhizal partnership. Indeed, they suggested that the evolution of plants and the colonization were tied to an association of an aquatic fungus with a semiaquatic alga. The fossil record would appear to support these ideas. Fossilized fungi have been documented from the Silurian contemporaneous with plant remains. The role and importance of fungi in the land colonization, frequently ignored, must be given more credence in future research.

3.9 Summary

This chapter considered the land migration and colonization by plants by first examining the key challenges of the land environment for organisms adapted to an aquatic habitat. These challenges include water and nutrient deficiency, problems of gas exchange, the need for mechanical support, and reproductive problems relating to fertilization and dissemination of propagules.

To effect survival in the new environment some major biochemical and morphological changes would have been required. The nature of these changes were such that they should have left four "footprints" in the fossil record: tracheids and thickened walls (conducting tissues and mechanical support), stomata and pores (gas exchange and transpiration), cuticle (waterproofing), and thick-walled spores (non-aqueous environment reproduction).

The fossil record does show all of these innovations and we believe it lends support also to the sequence we propose for their appearance. Of these four footprints, we believe that spores are the best evidence for earliest terrestrialization. We propose from these data and the time frame of the other footprints that the beginnings of terrestrialization occurred about 450–460 million years ago and that a vascular land flora (represented by the Rhyniopsida and cooksonioids) was established by 420 million years ago. The evidence—cytological, biochemical, and molecular—indicates that the nearest ancestors of the land flora are represented by modern day examples of charophycean green algae, such as *Coleochaete.*

The ambiguous fossil record of the possible ancestors suggests that there were numerous attempts and failures during this land migration. We thus caution that many fossil specimens that might be taken as transitional in some way may not represent actual ancestral lineages of the vascular plants, but parallel evolutionary experiments that failed early on. Then too, in all likelihood the vascular plants may be polyphyletic—just as the bryophytes now seem to be. Moreover, possible phylogenies that result from a consideration of all the data indicate that the bryophytes played no ancestral role in the evolution of vascular plants.

4

Fossil Evidence of Metazoan Transitions: Cambrian Through Middle Silurian

EVERETT C. OLSON

By the end of the Paleozoic both plants and animals had become firmly established on the continents and were diversifying into a wide range of environments. The roots of this exploitation of land may stretch back to the Cambrian and possibly into the Precambrian. Bacteria and protists have left almost no tangible terrestrial records, but these microscopic organisms surely were on the continents by Cambrian or earlier times. The earliest unequivocal fossils of terrestrial animals, however, are from rocks of latest Silurian and earliest Devonian age. For that time, as has been repeatedly stressed by Rolfe (e.g., 1980, 1985), the record is scant and incomplete. Even so there can be little doubt that the initiation of metazoan life on land well predated the late Silurian (Behrensmeyer et al. 1992). Future discoveries of fossils may give more concrete information on terrestrial migrations prior to the late Silurian. For the present, however, clues must come primarily from inferences based on marine organisms that lived in environments deemed suitable to potential invaders of the continents and from physical constraints imposed by those earliest continental habitats.

Some lineages surely passed through freshwater habitats in the course of transitions from strictly marine to strictly terrestrial existence. In a sense freshwater habitats served as "way stations" in a stepped process of land invasion. Many of the lineages, however, did not pass beyond the freshwater stage, becoming permanent inhabitants of this realm. Much of this stepped transition likely began early in the Paleozoic, but sampling of freshwater environments is biased and spotty and probably does not provide a true estimate of the extent of the process.

A persistent problem, noted at various places in this book, is why evidence of land invasions appears so late in the fossil record despite the long-time existence of seemingly appropriate physical conditions on the continents and the presence of possible invaders of land in marine and brackish environments. The absence of early evidence may be partly a function of the record, but this is certainly not the full answer.

The environmental constraints have been treated in chapter 2. By the Cambrian the oxygen (and shielding ozone) concentration was high enough and the carbon dioxide concentration (and its greenhouse effect) was apparently low enough to support life on land. Surely there were metazoans morphologically and physically able to occupy continental waters and at least marginal land environments. The primary constraint would thus appear to lie in the development of ecosystems sufficient to support complex metazoans beyond mere brief periods of occupancy or short forays onto land. The nature of such ecological constraints have been considered by many authors in recent times (e.g., Gray and Boucot 1977; Gray 1988; Gray and Shear 1992; Swain and Cooper-Driver 1984). Beerbower (1985) summarized his views on this problem as follows:

In terrestrial habitats, the limited availability of water as well as the slow release and low retention of organismic nutrient would have kept primary production to a very low level except at favored sites. High physical stress and uncertainty would have further limited production to small, scattered sites. The primary producers were predominantly microphytes, with macrophytes contributing very little if anything. Secondary production in the micro- and meiofaunas must also have been limited in amount and in temporal and spatial distribution. The fauna presumably included a mixture of microherbivores and decomposers (representing a second trophic level) plus meiodetritivores and meiopredators (the third trophic level). . . . The lack of fossils representing a macrofauna suggests the absence, or at least extreme rarity, of animals at the fourth level (except parasites). The typical structure of these early terrestrial ecosystems would thus have comprised three trophic levels with a decomposer loop.

Entrance of metazoans into such a system would presumably have been through transitional zones in salt, brackish, or fresh water circumstances. Whatever their course to land, the invaders would have need of such a minimal ecosystem base to persist for any extended time on land. The seemingly paradoxical delay in occupancy can thus most likely be explained in an ecological framework like this.

For reference throughout the rest of this book, figure 4.1 summarizes the sequence of major events in the terrestrial invasions of animals and table 4.1 gives a classification of the higher categories of metazoans that have successfully made invasions of the land. Nine phyla are included. In three of these phyla (Platyhelminthes, Nemertinea, Nematoda) all terrestrial representatives are considered cryptic. A fourth phylum, the Uniramia (including all the insects and related groups), today is by far the most numerous, diverse, and completely terrestrial of all, but it also lacks any fossil record relevant to the multiple water-land transitions that occurred in its various subgroups. The rest of this book omits further discussion of these four phyla because, as outlined in chapter 1, there is no evidence known that is relevant. We therefore focus on the remaining five: Mollusca, Annelida, Chelicerata, Crustacea, and Chordata.

The classification of arthropods used here follows Tiegs and Manton (1958), Manton (1977), and Schram (1978). This classification assumes a polyphyletic origin—that is, there is no single phylum "Arthropoda." The actual nature of arthropod origins remains unknown.

[Editorial note: The present authors are not systematists, and specifically are not specialists working with the arthropods. We recog-

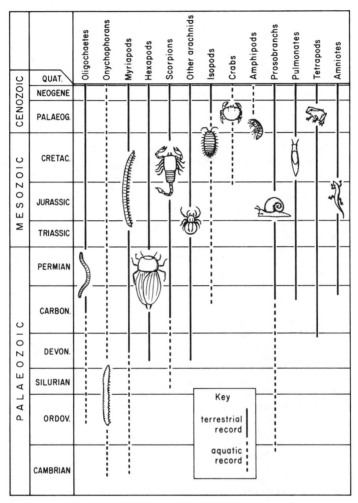

FIGURE 4.1 Approximate sequence of major events in the invasions of the land by animals, as known from the fossil record. Dotted lines indicate aquatic record; solid lines terrestrial record. *After Selden and Edwards 1989.*

nize that the subject areas of arthropod phylogeny, systematics, and evolutionary biology are active and rapidly changing. Recent authors (e.g., Kukalova-Peck 1990) have severely critiqued the views of Manton and associates and are developing both terminology and scenar-

TABLE 4.1
Classification of Animal Groups with Terrestrial
Representatives

PHYLUM PLATYHELMINTHES
  Class Turbellaria
    Order Tricladida C
PHYLUM NEMERTINEA
    Order Hoplonemertini C
PHYLUM NEMATODA C
PHYLUM MOLLUSCA
  Class Gastropoda
    Subclass Prosobranchia
      Order Archaeogastropoda L
      Order Mesogastropoda L
    Subclass Pulmonata
      Order Basommatophora P
      Order Stylommatophora P
PHYLUM ANNELIDA
  Class Polychaeta
    Order Phyllodocemorpha C
    Order Spiomorpha C
  Class Oligochaeta
    Order Moniligastrida C
    Order Haplotaxida C
  Class Hirudinea
    Order Rhynchobdellida L
    Order Arhynchobdellida L
(ARTHROPODA)
PHYLUM CHELICERATA
  Class Xiphosura A
  Class Eurypterida F, A
  Class Arachnida
    Order Scorpiones P
    Order Pseudoscorpiones P
    Order Opiliones P
    Order Acari C, P
    Order Palpigradi C
    Order Uropygia C (Thelyphonida)
    Order Schizomida C
    Order Araneida P
    Order Amblypygi C (Phrynichida)
    Order Solifugae P
    Order Ricinulei C
    Order Architarbi F, C
    Order Haptopoda F, C
    Order Trigonotarbi F, C
    Order Kustarachnae F, C
    Order Anthracomarti F, C

TABLE 4.1 (*Continued*)

PHYLUM CRUSTACEA
  Class Branchiopoda C
  Class Ostracoda C
  Class Copepoda C
  Class Malacostraca L, C, P
    Superorder Peracarida
      Order Isopoda L
        Suborder Oniscoidea P
      Order Amphipoda L
        Family Talitridae C, L
    Superorder Eucarida
      Order Decapoda L, C
        Suborder Brachyura L, C
          Family Ocypodidae L, C
          Family Gecarcinidae L, C
          Family Grapsidae L, C
        Suborder Astacidea L, C
          Family Astacidae L, C
          Family Parastacidae L, C
          Family Austroastacidae C
          Family Thalassinidae C
        Suborder Anomura L, C
          Family Coenobitidae L, C
PHYLUM UNIRAMIA
  Subphylum Onychophora C
  Subphylum Myriapoda
    Class Arthropleurida F, C
    Class Pauropoda C
    Class Symphyla C
    Class Chilopoda P
    Class Diplopoda C, P
  Subphylum Hexapoda
    Class Apterygota
      Order Thysanura C
      Order Diplura C
      Order Protura C
      Order Collembola C
      Order Monura F, C
    Class Pterygota
      Infraclass Palaeoptera P
      Infraclass Neoptera P
PHYLUM CHORDATA
  Subphylum Vertebrata
    Class Osteichthyes
      Subclass Actinopterygia L
      Subclass Sarcopterygia
        Order Crossopterygia
          Suborder Rhipidistia A?, F
        Order Dipnoi A?, C
    Class Amphibia
      Subclass Labyrinthodontia F
        Order Ichthyostegalia A

TABLE 4.1 (*Continued*)

Order Temnospondyli A?, L
Order Batrachosauria (Anthracosauria S. L.) A, L?
  Suborder Embolomeri A
  Suborder Seymouriamorpha A, P
  Suborder Diadectomorpha A, P
Subclass Lepospondyli* F
  Order Adelospondyli A
  Order Aistopoda A
  Order Microsauria A, L?
  Order Lysorophia A, C?
  Order Nectridia A
Subclass Lissamphibia
  Order Caudata A, L
  Order Gymnophiona A
  Order Anura A, L
Class Reptilia‡
  Subclass Anapsida
    Order Captorhinomorpha P
    Order Chelonia A, P
  Subclass Theropsida (Synapsida) P
  Subclass Sauropsida (Diapsida) P
Class Aves P
Class Mammalia P

A: amphibious. C: cryptic. F: fossils only. L: few terrestrial representatives, primarily aquatic or amphibious. P: predominantly terrestrial, but in some instances with aquatic forms derived from terrestrials.
*Lepospondyli is not a monophyletic group, and relationships of most orders are not well understood.
‡The monophyletism of Reptilia has been questioned. Also, Mammalia is sometimes classified with Theropsida (often termed Synapsida) to form a single class, and Aves has been considered a subgroup of Sauropsida.

ios substantially different from hers. We are unqualified to judge the specific issues and expect that additional changes will be proposed in the future. We use the Mantonian terminology here because it is widely familiar to nonspecialists. Most important, we are sure that neither the biological facts nor the interpretations we deal with here would need to be modified significantly because of future terminological or inferred phylogenetic changes.]

This chapter is devoted to the fossils and geology of Cambrian and Ordovician times, from which no terrestrial animals are known, and of the early Silurian, when the first traces have been found. Our treatment ends with the beginning of the Pridoli epoch of the late Silurian. The brief discussions of these periods and their geology in chapter 2, along with maps (figures 2.2–2.5) of the continents and oceans, will serve as the general geographic base to which additional detail will be added as necessary.

A time-stratigraphic organizational format is used in this chapter and in chapters 5 and 6. This framework makes it possible to integrate the geological, geographical, taxonomic, phylogenetic, and ecological aspects of the invasions of land into a coherent, temporal pattern. Each of these disciplines is important in itself and each has already been explored individually and in partial integration in the voluminous literature on the development of terrestrial animals and their exploitation of the continents. Many of these individual studies and comprehensive compilations are cited in the text. The following are particularly useful: Kevan, et al. (1975); Størmer (1976); Ziegler et al. (1979); Morell and Irving (1978); House et al. (1977); Manton (1977); Panchen (1980); Little (1983, 1990); Chaloner and Lawson (1985); and Selden and Edwards (1989).

4.1 The Cambrian

The beginning of the Cambrian is marked by the appearance of marine benthic and burrowing metazoans with well-formed skeletons, constituting the so-called Shelly Zone. A long-standing and still open question is the source or sources of this assemblage of hard-bodied organisms, which seems to appear abruptly. The sources and times of origins of the marine metazoans that were precursors of the lineages involved in later terrestrial invasions, while seemingly remote to the central questions of this book, are nevertheless important to understand.

New discoveries made during the 1970s and 1980s have greatly expanded the knowledge and understanding of the metazoans that arose before the Cambrian—especially those of the Vendian System, which includes the important Ediacaran faunas (Fedonkin 1985; Conway Morris 1987, 1989; Briggs and Fortey 1989; Seilacher 1989). Most or all of these Proterozoic metazoans were soft-bodied pelagic or epifaunal creatures. It remains unclear how, or whether, most of them were related to the well-known lineages of the Cambrian. Interpretations vary widely. More or less standard is the proposition that the progenitors of the prominent Cambrian phyla can be identified among the fossils of the Vendian. Seilacher (1989), however, argues that these organisms represent separate and failed evolutionary experiments and lineages.

Recent studies employing morphological, geological, and molecular data have suggested other relationships and phylogenies, including the possibility of polyphyletic origin of metazoans (Field et al. 1988; Valentine 1989; Lake 1990; Jacobs 1990). These studies bear on the ultimate origin and relationships of molluscs, arthropods, and chordates—groups that include the majority of metazoans that have made successful invasions of the land. As yet, however, no widely accepted solutions have been forthcoming.

During the Cambrian increasingly rich faunas of benthic invertebrates developed, including both epifaunal and infaunal elements. Phosphatic and phosphatic-calcareous skeletons common in the early Cambrian were partially superseded by calcitic structures. The advent of thicker external skeletons enhanced chances of fossilization, and to some extent gives a false impression of an increasing scope of radiation. Soft-bodied metazoans undoubtedly continued to be a significant, probably major, part of the Cambrian community, a speculation confirmed by several instances where unique preservation has allowed characterization (McMenamin and McMenamin 1990). No evidence of any terrestrial animals or of transitions from water to land has been found in Cambrian strata, although, as discussed in chapter 2, physical conditions suitable for terrestrial invasions seem to have existed.

By the late Cambrian, typical marine fossil assemblages occur in sediments deposited around the paleoequator and extending to about 50 degrees north and south latitudes (figure 2.3). Shallow seas bordered the continents and inundated significant portions of Laurentia, Siberia, Baltica, and Gondwana. The margins of these seas must have provided suitable avenues for exploitations of the land. Paleoclimatological reconstructions, based on climatologically sensitive sediments, latitudinal dispositions of the continents, and reconstructions of wind circulation patterns, indicate suitable temperatures and moisture. Several deposits, in particular, provide additional insights into more obscure Cambrian faunas, some of which may have bearing on possible progenitors of terrestrial organisms. Of these the Burgess Shale and the Wheeler and Spence Formations, all middle Cambrian, are of special interest. Less well studied sites from some thirty early, middle, and late Cambrian localities have provided additional important details.

4.1.1 The Burgess Shale Fauna

The Burgess Shale is a now-famous deposit of fine black shales of Cambrian age in the mountains of British Columbia. This deposit was discovered and described by Walcott (1911), who made extensive studies of the soft-bodied invertebrate animals that were exquisitely preserved. Whittington (1978, 1985) made extensive analyses and revisions of Walcott's work, and studies by Conway Morris (1982, 1989) and Conway Morris and Robison (1988) have added many details. Gould (1989) has presented an excellent description and somewhat controversial evolutionary interpretation of the faunas.

Many of the soft-bodied and weakly chitinized fossils are remains of otherwise unknown types of organisms, many not initially assignable to any recognized phyla. However, very recent reinvestigations have resulted in the accommodation of many of these enigmatic organisms within modern, higher-level taxa (Ramskold and Xianguang 1991; Butterfield 1990). The majority of assignable animals are arthropods. Burgess Shale specimens have added to the knowledge of the detailed anatomy of the phylum, but they have no direct relationship to the origins of land animals. A chordate, *Pikaia*, is also present in the Burgess Shale, but it has no direct relationship to the origins of land vertebrates.

One of the most significant animals of the Burgess Shale, for our purposes, is *Aysheaia pedunculata* Walcott (figure 4.2). Its importance as a possible progenitor of land animals, however, depends on its broad taxonomic relationship—which remains in dispute. The most exhaustive study has been made by Whittington (1978). He concluded that *Aysheaia* cannot be reliably placed in any recent higher taxon, but that it does appear to be a fossil lobopod (Snodgrass 1938), which would place *Aysheaia* in the Uniramia. The only living lobopods are the onychophoran uniramians, such as *Peripatus*. At the time of Whittington's work *Aysheaia* was the only known fossil lobopod of any age. A possible late Carboniferous representative from Mazon Creek, Illinois, (Thompson and Jones 1980) and an undisputed specimen from Montceau-les-Mines (Rolfe et al. 1982) have since augmented the record.

Manton (1977), in pressing the case for multiple origins of the arthropods, suggested that the uniramian type of appendage developed from a lobopodan ancestor in a lineage different from that of

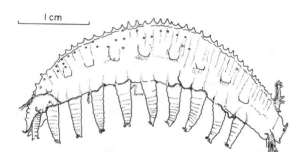

FIGURE 4.2 Reconstruction of *Aysheaia*, a presumed lobopod from the Cambrian. *After Whittington 1978.*

the biramian condition of the chelicerates. Under this concept the two major stocks of terrestrial arthropods, the uniramian insects and related groups and the chelicerate arachnids, were separated very early in the context of prearthropods. *Aysheaia*, interpreted as a lobopod, indicates the minimum antiquity of this separation.

In addition to the onychophorans, the tardigrada have been considered as a possible derivative of a lobopod such as *Aysheaia* (Cave and Simonetta 1976; Dzik and Krumbiegel 1989). The tardigrades (water bears) of the phylum Tardigrada are cryptic terrestrial animals with about four hundred extant species. Their known history dates back to the Cretaceous. If the lobopod/onychophoran relationships are accepted, a tangible base of one of the most important and the earliest groups of land animals was present in the Cambrian. Alternative relationships have been suggested by McKenzie (1983), who recognized possible affinities of *Aysheaia* to elapisoid echinoderms or polychaete annelids.

4.1.2 Other Similar Faunas

The assemblages of soft-bodied organisms first discovered in the Burgess Shale are now known from a number of sites. The most significant of these are in the Wheeler, Spence, and Marjum formations of the middle Cambrian in Utah (Conway Morris and Robison 1988; Robison 1985, 1990; Conway Morris 1990). The lobopod *Aysheaia* is known from a well-preserved specimen from the Wheeler Formation, where it occurs in association with a variety of other organisms that are also found in the Burgess shale. A similar suite occurs in the Spence and Marjum formations, where cyanobacteria, algae, and a variety of soft-bodied metazoans make up the assemblage. In addition, the Burgess Shale facies was widely developed around the Laurentian Craton (California, Idaho, British Columbia, Pennsylvania, and northern Greenland). Similarly, finds have been made in deposits associated with the Chinese Craton (early Cambrian) separated by deep, oceanic water from the Laurentian sites (Conway Morris and Robison 1988).

In a report on *Aysheaia* Robison (1985) assigned the genus to the subphylum Onychophora, but he notes that in his opinion any association with the Tardigrada is indeterminate. A myriapod-like fossil, *Cambrodus* (Robison 1990), from the middle Cambrian in Utah has a body form like that of the myriapods, but jaws are not present. *Xenusion* from the early Cambrian in Sweden is also relevant to assessments of the forerunners of terrestrial arthropod lineages. Two specimens have come from glacial erratics that were determined to be early Cambrian (Dzik and Krumbiegel 1989). Based on a reassessment by Dzik and Krumbiegel of the morphology of *Xenusion*, it appears that this Cambrian genus is related to both onychophorans and tardigrades.

Possibly bearing on the matter of Cambrian progenitors of terrestrial lineages is *Serracaris* (Briggs 1978) from the early Cambrian Kinzers formation of Pennsylvania. Only the anterior portions are well preserved. *Serracaris* resembles trilobites (which were all biramian), but it is distinct and has some uniramian characters suggestive of myriapods.

Combined, these tenuous leads are slowly building a tangible picture of the possible Cambrian predecessors of lineages within which transitions from water to land took place. On the basis of these data it appears likely that the earliest terrestrial metazoans may have been uniramian onychophorans and myriapods, both of which today are cryptic rather than fully terrestrial. Much is yet to be learned, but it does appear that continuing field and laboratory studies will yield increasingly important results.

A geological phenomenon in Sweden called "Orstens" also bear on efforts—if only tangentially—to find ancestral lineages. Over a broad area of Scånia and Västergutlan of Sweden and the Baltic island of Øland occur calcareous, whitish lenses in Orsten deposits of late Cambrian black shales (Müller 1983). Minute fossils of *crusta-*

ceans, including *ostracods* and several unassignable orders, occur in abundance. These kinds of assemblages and the genera that constitute them are unknown in Cambrian deposits anywhere else in the world. Like the animals of the Burgess-type shales, these unique organisms indicate the existence of a greater variety of Cambrian organisms beyond that recorded in beds containing typical hard-part fossils.

Ostracods have an excellent post-Cambrian record in both marine and freshwater deposits. Although most extant ostracod genera are marine, cryptic terrestrial representatives are known today, and they likely developed from several lineages that had diverged early in the Phanerozoic. They occur mostly in leaf litter and similar moist or wet environments. As they are protected only by a cuticle, they are unlikely to have been preserved as fossils. Much the same applies to several other crustacean groups.

Among the crustaceans only the oniscoidean isopods have attained full terrestriality and have radiated into several environments (Edney 1968). None of the Cambrian crustaceans of the Orstens have any characters suggesting more than very remote relationships to this suborder—or even to the parent order, Isopoda. What they do show is that in addition to well-known early Paleozoic animals, there existed in the Cambrian a variety of "primordial" crustaceans and that many possible sources of subsequent terrestrial occupancy were present. If, when, and how transitions may have been realized is unknown, but very unusual circumstances, such as those of the Orstens, give some hope that future finds among minute, obscure animals may eventually offer some insights.

4.1.3. Other Metazoan Groups

Members of the phylum *Chelicerata* are not known from the Cambrian. The xiphosuran-like Agalaspida were a widely distributed order of chelicerates, which appear in the fossil record throughout the Ordovician. This group includes four families, but it provides no clues to the origins of the eurypterids or arachnids, which figure importantly in incipient transitions to land during the Silurian and Devonian (Hesselbo 1992).

Annelids are represented in the Cambrian by marine polychaetes, but no direct evidence of the soft-bodied oligochaetes have been found. Burrows in marine sediments are common in the Cambrian, but the animals that formed them have not been determined. It is possible that some of the burrows were made by oligochaetes.

Gastropod molluscs, among which a number of subgroups have become terrestrial, enter the fossil record in the late Cambrian; Cambrian gastropods are assigned to the order Archaeogastropoda, which is significant to terrestriality only in representing primitive members of the class Gastropoda.

The earliest well-known *chordate* is the genus *Pikaia* from the Burgess Shale. *Pikaia* was first described as a polychaete worm by Walcott (1911) but was later assigned to the phylum Chordata by Conway Morris and Whittington (1979). Pteraspids, heterostracan agnathans (jawless fishes; Repetski 1978) and conodonts are the earliest known chordate vertebrates (Sansom et al. 1992). Both pteraspids and conodonts appear in the late Cambrian. The Calicichordata (Jeffries 1986) have been assigned to the chordates, but many echinoderm experts consider them echinoderms. Graptolites, another enigmatic group, occur in the Cambrian and also have been considered to be chordates, but the assignment is open to question.

The conodonts were first recognized in the middle of the nineteenth century and have had a checkered systematic career ever since—including assignments to algae, various group of multicellular plants, and a variety of major metazoan classes. They are best known from very small, hard structures of extremely varied shapes composed of the mineral apatite. Few are greater than 2 mm in their major dimension. Many are toothlike (hence the name), but others resemble rods, bars, and plates. They range from Cambrian to Triassic, occurring as single elements and occasional complex assemblages.

The conodont body consisted of soft tissue and only very recently has been identified in the fossil record. Very few examples of the "conodont animal" have been found. From some of these specimens and the presence of apatite in the hard parts a long-standing opinion that conodonts had chordate affinities was confirmed (Briggs et al. 1983; Sansom et al. 1992). Moreover, detailed histological studies of some early conodonts have now produced unequivocal evidence of the presence of cellular bone, a definite vertebrate character. Thus at least some of the Cambrian conodonts are vertebrates, and they were clearly different from the other Cambrian vertebrates, the agnathans. It is now certain that at least two groups of vertebrates had come into being by late Cambrian times.

4.2 The Ordovician

The Ordovician period appears to have been eminently suitable for penetration of freshwater and land habitats by metazoans, but direct evidence that colonization began as early as the Ordovician is meager. The continents, especially Laurentia, had retained favorable (tropical or temperate) positions (figure 2.4), although Gondwana's southward movement had by then taken it partly into the south polar zone (where it developed glaciation late in the period). Broad continental expanses, however, persisted in the circumtropical areas, and large, shallow epicontinental seas were widespread. Presumably the seas were bordered by lowlands, estuaries, and lagoons.

Planktonic and benthic *marine* invertebrates flourished and included additional taxa poorly represented or unknown in the Cambrian. Arachnids, merostomes, and agnathans are all well represented in the Ordovician. Arachnids and merostomes may be related to terrestrial migrants. Some of the fossils occur in beds which may have been formed in brackish waters, and hence the animals from which they came likely had developed rather broad osmoregulatory tolerances.

Fossil marine worms from the Ordovician have been assigned to the oligochaetes. The remains appear undiagnosable and the presence or absence of marine oligochaetes during the Ordovician is an open question.

4.2.1 Continental Deposits

Continental deposits are known only in the form of ancient soils, paleosols, (Retallack 1981, 1985) and braided stream deposits (Cotter 1978). Mixed oceanic-continental deposits and possibly fluvial sediments characterize at least part of the Harding sandstone of the western United States. This sandstone is a primary source of possible candidates for land migration. Using data from this formation as a mainstay, Beerbower (1985) has attempted to assess the nature of development of a continental ecosystem in the Ordovician and has speculated that radiation into freshwater lacustrine habitats is compatible with the evidence but not subject to validation, in view of the sparsity of continental deposits.

4.2.2 Mixed Oceanic-Continental Deposits

The sediments of the Harding sandstone form a complex of beds deposited under a variety of marine, subtidal, and supratidal environments. In individual sections they vary from the base to the top, as a rule passing from moderate to shallow marine environments in the sequence. They also vary geographically over the areas of exposure in Colorado, Wyoming, South Dakota, and Montana. Of greatest interest here are the animals in the deposits, of which the remains of agnathan fishes have received the most attention. These fossils, until recently the oldest known vertebrates, have directed attention to the Harding sandstone ever since their first description by Walcott in 1892. The nature of the habitat of the agnathans received special attention because it had bearing on the long-lasting problem of the marine, estuarine, or freshwater origin of vertebrates (Romer and Grove 1934; Denison 1956; Gans 1989).

The current consensus is that these deposits were all formed under marine conditions, partly in the neritic zone and partly near shore; there is, however, evidence to the contrary (Graffin 1992). Deposition took place in relatively quiet waters (Denison 1956), interpreted as perhaps lagoonal by Spjeldnaes (1979). Fisher (1978), however, inferred an estuarine environment, with moderately dynamic conditions that gave rise to a mixed zone of point bars and channel deposits that graded into mudflats. Sedimentary analyses of a locality in Colorado where vertebrate scraps are abundant (Graffin 1992) suggest that the bone-bearing facies were formed under fluvial, freshwater circumstances and that the vertebrates lived in freshwater streams and rivers. Such differences of opinion characteristically plague interpretations of the highly varied beds of mixed oceanic-continental deposits, for lateral and vertical variation often is rapid; even nearly adjacent sections may suggest quite different interpretations.

Living in a near-shore environment were animals of an ecosystem including merostomes, arachnids, and trilobites (the latter, relatively rare). Trace fossils have been particularly important for reconstructing the taphonomy of this assemblage (Fisher 1978). Bioturbation is extensive, indicating a well-developed infaunal complement. Conodonts, linguloid brachiopods, bivalves, gastropods, and cephalopods are present in various sites in the Harding sandstone, but of greatest importance with regard to transitions to land are the arachnids and merostomes, groups which include the very earliest known terrestrial animals.

The littoral and sublittoral environments, regardless of the precise interpretation of the environments of deposition, afforded excellent opportunities for invasion of land. The area of deposition of the Harding sandstone lay along and to the north and south of the Ordovician paleoequator, and its geographic position with reference to the land mass of Laurentia was such that the probable wind directions suggest a warm and moist climate.

Agnathan vertebrates of the Ordovician were not direct progenitors of the gnathostomes, from which land-adapted lineages eventually developed. Agnathans, therefore, probably could not have made any significant penetration of the land. On the other hand, the merostomes and arachnids surely could have penetrated beyond the tidal zone, if only in brief forays. Boucot and Janis (1983) and Beerbower (1985) suggested the likelihood that merostomes and arachnids lived partly in brackish waters (especially under Fisher's interpretation). This would indicate a tolerance of varying salinities and would make these two groups candidates for eventually invading freshwater, as appears true for the agnathans. Indications from early Devonian deposits that have produced land animals are that the early ecosystems included primarily minute metazoan animals. No direct predecessors of these tiny metazoans are known from the Harding sandstone.

4.2.3 Paleosols

Late Ordovician paleosols exist in Nova Scotia and in the eastern United States in the Potters Mills Clay (Retallack 1981, 1985, 1986). Throughout the fossil record from the Precambrian Huronian to the Pleistocene, paleosols provide opportunities for examinations of events often not otherwise recorded. Some formed on top of major unconformities, where they represent a shift from an erosional to a depositional environment. The Nova Scotia paleosols, for example, rest on volcanic rock.

Beyond giving some information on the existence of terrestrial plants, and possibly protistans, paleosols have little direct bearing on metazoan land invasions. The Potters Mills Clay, however, is marked by abundant burrows (figure 2.1), which indicate the presence of fairly large cryptozoans, possibly millipedes (Retallack and Feakes 1987). The identity of the animals that made the burrows necessarily is obscure. No actual remains of organisms have been found. What is shown, even without clear assignment, is that fairly large metazoans had migrated onto the continent by the late Ordovician. A more complex ecosystem than commonly envisaged may thus have been in existence (Behrensmeyer et al. 1992). No plant roots have been identified in Ordovician paleosols, although they are present in later paleosols.

4.3. The Silurian (Pre-Pridoli)

The Ordovician-Silurian boundary has been defined on the basis of a marine section in western England and Wales in which the rocks of the two periods are separated by a marked unconformity. At this time the glaciation in the Antarctic regions of Gondwana, while still present, was waning. The result was that sea level initially was low relative to the continents and marine faunal provincialism was widespread. During the Llandovery epoch (early Silurian), with the continued melting of the ice, epeiric seas became widespread and cosmopolitan marine faunas developed. As shown in figure 2.5, by the middle Silurian (Wenlock), Laurentia and Baltica lay astride the equator and extended about 30 degrees to the north and south. These continents have been the principal sources of animals that bear relationships to later terrestrial invaders.

4.3.1 The Sedimentary Record

Sedimentary deposits in the widespread epeiric seas were in large part autochthonous (Ziegler et al. 1979) and hence are good climatic indicators. The sediments, along with reconstructions of continental topography and of wind directions, indicate that climates in the circumequatorial region were wet and hot, whereas from about 10 degrees to 30 degrees north and south they were warm and dry. Conditions appeared excellent for occupancy of land.

Relative movements of Laurentia and Baltica continued the reduction of the Iapetus Sea that had been evident during the Ordovician, and by late Silurian the sea was closed. The accompanying Caledonian orogeny had produced active areas of uplift, expressed in extensive mountainous uplands along the general zone of contact. Sediments derived from these highlands produced the Old Red Sandstone, which spans the Silurian-Devonian boundary and which (in Wales and Scotland) has produced the earliest remains of clearly defined terrestrial metazoans, diplopod millipedes. This sandstone has the greatest promise of yielding information on the progressive occupation of land.

The boundary between the Silurian and Devonian is, however, much less distinct than that of the Ordovician-Silurian. The Pridoli epoch of the late Silurian and the earliest Devonian rocks form a more or less continuous depositional series in the areas of the Old Red Sandstone. For this reason, as well as the fact that the first uncontested terrestrial metazoans occur in the Pridoli time span, the uppermost Silurian will be treated with the Devonian in the next chapter. Here we will be concerned with the pre-Pridoli Silurian.

The rich marine invertebrate faunas that lived in the epeiric Silurian seas contrast sharply with the limited and restricted faunas found along the margins of those seas. It is in the latter sediments, formed under near-shore and mixed marine-continental environments, that the precursors of land animals must have lived and from which the very meager remains of these precursors have come.

4.3.2 Near-Shore Faunas

Eurypterids, although by no means common, are the most abundant and best preserved of the animals that have some relevance to the origins of land animals. None of them became fully terrestrial, but among them were animals apparently adapted to shallow-water marine and brackish environments. Inferred "walking" patterns based on trace fossils similarly point to a wide range of habits. Some of the walking patterns suggest animals that moved about only partly covered by or even free of water (Briggs and Rolfe 1983; Selden 1981, 1984). Tracks from four temporally separated sites from the Silurian to Carboniferous suggest a grade progression from swimming and out-of-phase walking to in-phase walking, perhaps involving an increasing development of partial terrestriality. The respiratory structures, as far as they can be determined from the fossils, were incapable of gas exchange out of an aquatic medium.

Scorpions were almost certainly derived from eurypterid-like ancestors, from which they can be distinguished among fossils by the presence of pedipalpal chelae and pectines. The presence of the characteristic sting is difficult to determine in the earliest fossils of scorpions. Altogether eight specimens of pre-Devonian scorpions are known (Rolfe 1985; Kjellesvig-Waering 1986), and these are highly varied, indicating several lines of descent. The earliest known scorpion is Llandoverian in age and, like all pre-Devonian forms, was aquatic in habitat. The structure of the gills is unknown, but none of the specimens appears to have had a preoral chamber critical to feeding out of water (Størmer 1976). Although at one time scorpions were taken to be the earliest of the land animals, the myriapods and non-scorpionid arachnids, in fact, have produced the most ancient records known to date.

A few specimens of supposed pre-Pridoli *millipedes* and possible trace fossils have been described from Llandovery, Wenlock, and Ludlow beds. None of these, however, can be assigned to this group with certainty. *Necrogammarus salweyi*, from the Ludlovian, may be a very large millipede, but key defining features are not preserved. An undescribed but illustrated specimen (Rolfe 1980) from the Wenlock-Ludlow Fish Bed formation is perhaps a diplopod and possibly a millipede. *Archidesmus loganensis*, once described as a flat-back millipede, suggesting terrestrial habits, is indeterminate.

Xiphosurans occur throughout the Silurian in marine beds. The earliest complete specimen is known from Llandoverian beds in Wisconsin (Mikulic et al. 1985). All Silurian specimens thus far have indicated xiphosurans as being basically aquatic; but like later known Carboniferous and also living xiphosurans these earliest representatives of the lineage may have made short-term forays onto land. The Wisconsin Llandoverian site, however, while formed in a restricted environment, is marine and contains other arthropods including trilobites, phyllocarids, and ostracods, plus a possible marine uniramian and remains suggesting the presence of a leech which, if authentic, is the first known from the Paleozoic.

Significant Ludlovian trace fossils from Sweden include fecal pellets containing ascomycetous fungal hyphae (Sherwood-Pike and Gray 1985), suggesting the presence of fungivorous microarthropods (Rolfe 1985). Land occupancy by arthropods at this time is virtually confirmed by this occurrence (Gray and Shear 1992).

Agnathan vertebrates and the first traces of osteichthyan fish (acanthodians) occur in the pre-Pridoli Silurian in scattered localities. Remains have come from a variety of environments, mostly restricted marine, and some agnathans and acanthodians may have lived in brackish to freshwater habitats. None, however, has any close relationship to the early Devonian dipnoan and crossopterygian fishes, which are close relatives of the vertebrate invaders of land. No remains of either of these groups have been found in the Silurian.

4.4 Summary

Cambrian Of the currently known Cambrian fossils (all marine) the lobopodan *Aysheaia*, the possible myriapod *Cambrodus*, and the possible tardigrade precursor *Xenusion* provide information on the origin of some later terrestrial stocks. If *Aysheaia* and *Xenusion* are relatives of the Onychophora, today a cryptic terrestrial group, they suggest the existence of a soft-bodied marine source for the onychophorans along the lines suggested by Manton (1977). The only other fossil representatives of onychophorans are those from the late Carboniferous of Mazon Creek and Montceau-les-Mines. *Cambrodus*, of the middle Cambrian, indicates a marine relative of the earliest (Silurian) known metazoan invaders of land, which were also arthropods. Geological considerations indicate that land invasions could have taken place during the Cambrian, and a variety of marine organisms in shallow seas provided a reservoir from which terrestrial animals could have developed. There is, however, no evidence as of today that any invasions in fact occurred. Likewise, there is no clear evidence that a suitable trophic base for animals existed on land.

Ordovician The Ordovician for the most part has produced little tangible evidence of terrestrial metazoans or of transitional forms between the water and the land. The animals that formed the burrows of the Potters Mills Clays are possible exceptions. Both the biological and physical conditions, as shown by the sediments and animals of the Harding sandstone and the position and configuration of Laurentia, indicate that the stage for terrestrial invasion was set. Whether unrecognized biological constraints to land invasions, accidents of preservation, or insufficient sampling are responsible for the lack of evidence cannot be determined at present. Nevertheless, there seems reason to hope that one day additional remains of freshwater or truly terrestrial animals may be found.

Silurian (pre-Pridoli) As this brief survey shows, the only sound evidence of Silurian land occupancy prior to the Pridoli is the Swedish fecal pellets containing fungal remains, which indicate the presence of microarthropods. The arthropods as a whole have some points of interest with reference to land invasions. Perhaps the most significant of these is that the best-known groups other than trilobites—the eurypterids and xiphosurans—are characterized by a wide variety of animals, suggesting several lines of descent and the potential of several of them accomplishing temporary forays into marginal environments. This general pattern may carry over into other, ultimately more successful, groups. The several kinds of scorpions, for example, suggest such a pattern, although on the basis of minimal evidence.

The continents of the time were suitable for occupancy by land animals; development of soils indicate the existence of environments favorable to, and apparently occupied by, cryptic microarthropods. Whether or not invasions by macroscopic arthropods or other animals in fact took place cannot at present be determined, but it seems probable that at least the early stages of basic food pyramids were being developed on land.

5

Fossil Evidence of Metazoan Transitions: Late Silurian and Devonian

EVERETT C. OLSON

Invasions of land by metazoans were almost certainly well underway before the late Silurian. As noted in chapter 4, however, the evidence for this is equivocal. As far as the fossil record shows, it is the sediments of the latest Silurian—the Pridoli epoch—that have yielded incontrovertibly terrestrial fossils of metazoans. Even at this time, however, the record is extremely spotty, and both the organisms and the life environments are subject to various interpretations. In contrast, the record of vascular plants is clear and extends back into pre-Pridoli times (chapter 3). During the Devonian the plant cover of the continents increased, providing a progressively more diversified array of environments for terrestrial animals; forest cover was present in some areas by the late Devonian.

After the Devonian, exploitation of terrestrial environments expanded rapidly among both vascular plants and metazoans. Evolution of some of the principal terrestrial animals—in particular, pterygote hexapods (winged insects)—took place on land and not in the water (Behrensmeyer et al. 1992). In contrast, however, was the more gradual passage from water to land of the vertebrate tetrapods, already well under way by the latest Devonian. Full realization of terrestriality of both amphibians and reptiles took place during the latest Carboniferous and into the Permian (Olson 1984).

5.1 The Old Red Sandstone

Deposits of the Lower Old Red Sandstone formation—with outcrops ranging from northern Europe to eastern North America—have provided the most critical evidence of early transitions to land during the late Silurian and early Devonian. In these deposits are records of transitional zones between water and land formed under conditions suitable for the preservation of fossils. The Lower Old Red clastics include detrital sediments deposited under near-shore, marine, estuarine and lagoonal, freshwater, brackish, and fully terrestrial environments. As in all such sedimentary circumstances, the state of preservation of fossils varies widely, generally being fair to good in finer deposits (such as those formed in lakes, lagoons, and shallow marine sites) and poorer in coarser clastics formed under higher energy circumstances.

The relative strength and composition of the skeletal materials were also major factors in the state of preservation. Such differences in the preservability of taxa introduce severe biases into estimates of the composition of the biotas from which the samples are drawn. The extent of preservation similarly is crucial to taxonomic determination of groups of organisms for which precise placement is possible only if certain, often poorly preserved, features are present (Behrensmeyer et al. 1992).

The deposition of Old Red Sandstone was a consequence of mountain building during the Caledonian orogeny, which began in the late Silurian. It affected the British Isles, providing sediments formed in the Welsh and Scottish areas and on through Norway and Sweden, Spitzbergen, Greenland, and northeastern North America. By the middle Devonian the Acadian orogeny had commenced, initiating changes in the Appalachian area and producing extensive red beds in eastern Canada, New York, and Pennsylvania. Devonian highlands in the more western parts of North America produced sediments in which fossils, in particular of fish, were well preserved.

The Caledonian and Acadian orogenies and consequent tectonic and sedimentary adjustments resulted from the collision of Laurentia and Baltica to form a single continent, commonly termed Laurussia. This collision continued trends of movement initiated during the Silurian (figure 2.6). During the Silurian and Devonian as well, the Uralian margin of Baltica became active as a zone of compression, forming highlands in the general Uralian region (Ziegler et al. 1979; Ivanov et al. 1975).

By the middle Devonian much of Laurussia lay north of the equator, extending, however, from about 10 degrees south to 50 degrees north. Earlier, major portions of this "Old Red Continent" lay astride the equator, in tropical and subtropical zones. The temperatures during the early phases of deposition of the Old Red Sandstone appear to have been warm to hot, and rainfall (while substantial) seems to have been somewhat seasonal in the areas in which fossil remains of terrestrial plants and animals were preserved. Distributions and conditions of preservation of early vascular plants suggest that a major concentration of living plant assemblages existed around rivers, lakes, estuaries, and swamps (Edwards 1980). These moist settings presumably were habitats of the early terrestrial animals and provided pathways for the evolution of fully terrestrial forms.

The times and circumstances of development of freshwater and terrestrial groups among the vertebrates and invertebrates were disjunct, running somewhat parallel courses, but sufficiently offset in time that the potential environments presented quite different challenges and opportunities to transitional members. Representing possible progenitors of amphibious vertebrates were lungfishes (dipnoans) and lobe-finned fishes (rhipidistian crossopterygians), both of which were newly evolved members of the osteichthyans (bony fishes) and which make a first appearance in strata of the early Devonian. The primary invertebrate invaders for which there is a record were arthropods. Tangible evidence of interactions, such as predator-prey relationships, between the semiterrestrial or terrestrial invertebrates and the vertebrates has not been found in any deposits before the late Carboniferous, but it seems indisputable that such interactions actually did occur much earlier.

Nevertheless, because of the partial independence of the emergence of major invertebrate and vertebrate animals, the two groups are treated separately in the rest of this chapter. The interfaces and beginnings of ecological interactions are, however, an important part of our discussion of the Carboniferous exploitations of land, presented in chapter 6.

5.2 Invertebrates

Sporadic finds of fossils in pre-Pridoli Silurian beds give no clear evidence of the development of terrestrial invertebrates. The several finds of scorpions that at one time or another had been considered as terrestrial have now been interpreted as aquatic. The animals had not developed a preoral cavity, considered by Størmer (1976) to be a crucial step in the development of terrestrial feeding. The common eurypterids, as well, do not appear to have been capable of sustaining terrestrial life. Størmer's general interpretation of arthropods' partially and fully successful invasions of land are illustrated in figure 5.1.

The record from deposits of Pridoli age is somewhat better. Discovery of terrestrial arthropods—an arachnid trigonotarbid and two centipedes—from the basal Pridoli deposits (411 mya; Jeram et al. 1990; Rolfe 1990) has given direct evidence of metazoan occupancy of land by that time. The fragmentary remains occur in beds formed in an intertidal lag zone. The remains include those of fishes, eurypterids, aquatic scorpions, kampecarid myriapods, some restricted marine invertebrates, and land plants including *Cooksonia*. The terrestrial arthropods are predators, indicating that a fairly complex terrestrial ecosystem was well developed and that such systems likely had an extensive prior history. Additionally, a specimen from the Stonehaven mudstone in Scotland has been confirmed as a true millipede. The bed in which it occurs is of lacustrine origin; whether the organism was in fact strictly terrestrial or whether it was aquatic is indeterminate. In either case it is the first fully authenticated myriapod.

From this time as well, from Stonehaven and also from another Scottish site near Oban on the Island of Kerrara, have come poorly preserved remains of elongated, segmented animals that could be millipedes, but which cannot be placed with confidence except as Arthropoda *incertae sedis* (Almond 1985). The definitive features of myriapods are not preserved in these fossils. The generic name *Kam-*

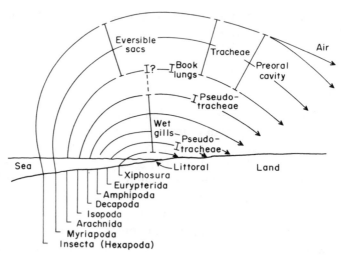

FIGURE 5.1 Interpretations of partially to fully successful invasions of land by various arthropods during the middle Paleozoic. The crucial respiratory features are indicated. *After Størmer 1977.*

			British Isles
Devonian	Famennian		
	Frasnian	Escuminac fm., Canada: ? *Arthropleura*	
	Givetian	Gilboa fauna, U.S.A.: centipedes	
	Eifelian		
	Emsian	Alken fauna, Germany: *Eoarthropleura*	
	Pragian	Rhynie Chert fauna : no Myriapoda *Kampecaris tuberculata*	
	Lochkovian	Arbuthnott Gp: *K. forfarensis, A. macnicoli* *Kampecaris dinmorensis* Oban fauna: *K. forfarensis,* ? *K. obanensis,* '*Archidesmus*' sp.	
Silurian	Pridoli	*Kampecaris obanensis,* Kerrara; undescribed kampecarids Stonehaven Gp: prob. diplopod, '*Kampecaris?*', and others	
	Ludlow	*Necrogammarus*	
	Wenlock	Undescribed 'Myriapods' (probably invalid)	
	Llandovery	*Archidesmus loganensis* (invalid)	

FIGURE 5.2 Summary of records of claimed Silurian–Devonian Myriapoda, arranged in stratigraphic order. *After Almond 1985.*

pecaris has been applied to various remains, which probably do not form a single genus but which are associated in the Kampecaridae from the Pridoli and early Devonian. All have some millipede features, but, as interpreted by Almond, are best considered more broadly as uniramians, perhaps closely related to the Diplopoda (see figure 5.2).

In post-Pridoli sediments of the Devonian, three deposits—the Rhynie Chert of Scotland, the Alken sediments of the Mosel Valley of Germany, and the black mudstones of Gilboa, New York—provide the great bulk of the data on probable terrestrial invertebrates. Other sites have as yet yielded little. But a few, such as the upper Devonian of Escauminac Bay, Nova Scotia, have the potential of producing important information (Labandeira et al. 1988; Shear and Kukalova-Peck 1990).

5.2.1 The Rhynie Chert

The Rhynie Chert member of the Lower Old Red Sandstone formation in Aberdeenshire, Scotland, is deservedly famous for its contribution to knowledge of early vascular plants: in it can be found *Rhynia, Cooksonia,* and *Asteroxylon* (figures 3.5 and 3.6). The Rhynie Chert also supplies glimpses of early terrestrial invertebrates, which along with the plants give insights into the ecology of the Devonian biota (Hirst and Maulik 1926; Kevan et al. 1975; Behrensmeyer et al. 1992; Gray and Shear 1992). The Rhynie deposits formed in an extensive peat bog, with the peat interbedded with layered siliceous beds that may have resulted from flooding episodes, perhaps fed by hot springs. The Rhynie Chert may therefore represent a very unusual depositional type.

Preservation of the small animal fossils is remarkably good, and some of the organisms are so similar to living counterparts that Crowson (1970, 1985) suggested that some might be the result of contamination. Crowson's interpretation has not been widely accepted, as the continuity of preservation of the animals and chert argues to the contrary (Whalley and Jarzembowski 1981; Greenslade 1988). The precise age of the Rhynie deposits has been debated, but it is generally considered to be Pragian (Westoll 1977), partly on the basis of spores. Størmer (1977) considered the age to be Emsian.

The Rhynie fauna, predominantly composed of microarthropods,

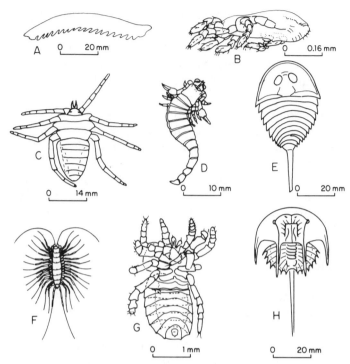

FIGURE 5.3 Paleozoic Onychophora and Chelicerata (Merostomata and Arachnida). Not drawn to scale. A: *Helenopdora*, a probable onychophoran from Mazon Creek, late Carboniferous. *Outline sketch based on Thompson and Jones 1980*. B: *Protacaris*, an arachnid mite from the early Devonian Rhynie Chert. *After Hirst 1922*. C: *Architarbus*, an architarbid arachnid from Mazon Creek, late Carboniferous. *After Petrunkevitch in Moore 1955*. D: *Palaeophonus*, a scorpion arachnid from the Silurian. *After Petrunkevitch in Moore 1955*. E: *Glypharthrus*, an Ordovician aglaspid merostome. *After Størmer in Moore 1955*. F: *Latzelia*, a centipede arachnid from Mazon Creek, late Carboniferous. *After Mundel 1979*. G: *Palaeocharinoides*, a trigonotarbid arachnid from the Rhynie Chert, early Devonian. *After Petrunkevitch in Moore 1955*. H: *Euproops*, a xiphosuran merostome from Mazon Creek, late Carboniferous. *After Fisher 1979*.

includes a collembolan apterygote (*Rhyniella*), a mite (*Protacaris*), and two trigonotarbid arachnids (*Palaeocharinus* and *Palaeocharinoides*). (See figures 5.3 and 5.4.) The "giants" of the fauna are the trigonotarbids, with some specimens having a body length of up to 14 mm (Rolfe 1980). Thin sections of these arachnids show the presence of

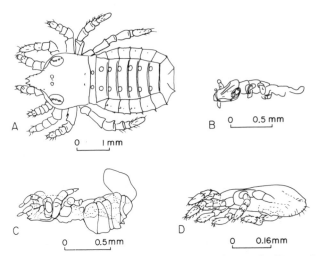

FIGURE 5.4 Representative metazoans from the early Devonian Rhynie Chert. (Not drawn to scale.) A: *Palaeocharinus*, trigonotarbid. B: *Rhyniella*, collembolan. C: *Palaeoctenzia*, possible spider. D: *Protacaris*, mite. *A, C, D after Hirst 1922; B after Hirst and Maulik 1925.*

book lungs and thus provide unequivocal evidence of the existence of fully terrestrial arthropods at this time. The collembolans were probably herbivores, but the predominant arthropods were carnivores. Their prey, which may have included the collembolans and mites, is largely uncertain and likely consisted of unknown elements of the system. Much of the food chain—probably based on algae, fungi, bacteria, and soft-bodied protists—has not been preserved. Kevan et al. (1975) and Rolfe (1980, 1985) have presented analyses of the paleoecology of these deposits from which our own conclusions are drawn; their works should be consulted for details, especially to distinguish what is firmly known from what is speculative.

5.2.2 The Alken Site

Near the village of Alken in the Mosel Valley of Germany occur sediments that contain a wide spectrum of marine, brackish, and freshwater invertebrates, as well as plants. The fossils are preserved in a black shale. Although less well preserved than those of the Rhynie Chert, the Alken fossils are more diverse. The preserved

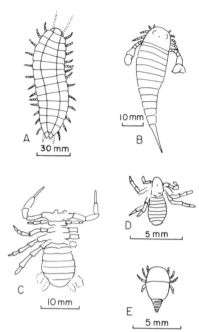

FIGURE 5.5 Examples of Alken fauna. (Not drawn to scale.) A: *Eoarthropleura*, an arthropleuran myriapod. B: *Parahughmilleria*, a centipede. C: *Waeringoscorpio*, a scorpion. D: *Alkenia*, a trigonotarbid. E: *Diploaspis*, a xiphosuran. *After Størmer 1970–1976, 1977.*

plants are largely vascular psilophytes, although algae also are present. Størmer (1976) interpreted the environment of deposition as "lagoonal," formed in water surrounded by vascular plants in a mangrove-like community, with the plants living a semiaquatic existence. Edwards (1980), however, has disputed this, indicating that the plant remains suggested growth under drier conditions, from which they were washed into a lagoonal environment.

The fauna of the Alken site is a mixed array of arthropods, bivalves, and fish. The arthropods include ostracods, eurypterids of several types, the xiphosurans *Diploaspis* and *Heteroaspis*, the arthropleuran myriapod *Eoarthropleura*, the scorpion *Waeringoscorpio*, and two trigonotarbid arachnids, *Alkenia* and *Archaeomartus* (figure 5.5). Størmer (1977) noted a possible spider, but the identification is uncertain. Among the arthropods, the arthropleuran myriapod and the trigonotarbids are clearly terrestrial, but Størmer (1976) concluded that the scorpion was aquatic because of the absence of a preoral cavity.

The matter of the eurypterids and xiphosurans is less clear. Some eurypterids had swimming legs and were clearly aquatic; others had walking legs useful for moving across aquatic substrate, but they may have made brief forays onto land. Størmer (1977) suggested on the basis of plate-shaped abdominal appendages that the eurypterid gills may have been temporarily shielded from desiccation. Xiphosurans, judging from the habits of *Limulus* and from interpretations by Fisher (1979) of late Carboniferous forms, may have spent part of their time out of water. In both instances the visits these organisms may have made onto land were unimportant from an evolutionary perspective, as neither lineage went on to develop fully terrestrial forms.

The Alken fauna, except for the trigonotarbids, is very different from that of the Rhynie Chert, although there is a general floral resemblance. Overall, the Alken fauna reflects a different pattern of land exploitation. Along with the Gilboa assemblage treated next, it illustrates, on the one hand, the probable complexity of land invasions during the Devonian and, on the other, how little tangible evidence for a total picture is now available.

5.2.3 The Gilboa Site

Table 5.1 summarizes much of the information from the Rhynie and Alken sites in Europe in comparison with the Gilboa site in the state of New York. The age of the Gilboa deposit is Givetian (middle Devonian) and thus considerably younger than the Emsian/Pragian deposits of Rhynie and Alken. The fossils occur in a black mudstone (or shale) and were carried by wind or running water into the water from adjacent environments. The fauna is relatively rich in remains of terrestrial arthropods (Shear et al. 1984). It includes remains of several arachnids (including oribatid and alicorhagid mites), trigonotarbids, a spider (Shear et al. 1989), two types of terrestrial centipedes, and a probable bristletail insect (based on a part of a head capsule).

TABLE 5.1
Devonian Fossil Localities

Locality and Age	Aquatic or Amphibious Faunas	Terrestrial Faunas and Floras	Habitat
Alken, Lower Emsian	Eurypterids, scorpion, possible merostome	2 genera of trigonotarbids, indet. arthropod, 1 arthropleurid	Shallow, brackish lagoon
Rhynie, Upper Emsian	Lepidocarid crustacean, possible merostome	3 species trigonotarbids, 1 mite, 2 collembolans, algae, lycopsids	Bog or lake margin, hot silica flooding
Gilboa, Givetian	Probable eurypterid	2+ trigonotarbids, 4 mites, possible amblypygid, 1 spider, 1 centipede, 1 pseudoscorpion, 1 arthropleurid, progymnospermopsids, lycopsids	Black mudstone, delta deposit; material transported probably short distance

SOURCE: After Shear et al. 1984 and Shear 1990.

The preservation is unique, and the remains are largely fragments of cuticle that appear to have been trapped in mats of stems of the lycopod *Leclerquia*. The black mudstones containing the fossils were treated with hydrofluoric acid; the residual fragments of cuticle were picked from the remaining organic material. Among the hundreds of specimens the mites are largely intact, and trigonotarbids are represented by partial individuals. The close resemblance of some of the animals to living counterparts, as in the case of the Rhynie Chert, raised the question of contamination, but this possibility was carefully investigated and eliminated (Kethley et al. 1989).

After the Givetian, throughout the remainder of the Devonian, no deposits with assemblages of terrestrial invertebrates have been found. Occasional fragments have been reported, but they do not add appreciably to understanding of the development of land invasions.

5.3 Vertebrates

Arthropods and vertebrates have been the principal exploiters of the terrestrial way of life, with molluscs and annelids occupying extensive but more restricted environments. Scant as the remains of inver-

tebrates are in rocks of Devonian age and earlier, they give a better
documentation of the stages leading to occupancy of land than do
those of the vertebrates.

5.3.1 Evidence of Tetrapods

Dipnoan fishes, an order of bony fishes distinguished by their having
lungs for air breathing, have made abortive transitions to temporarily
nonaquatic circumstances by surviving dry seasons in states of aesti-
vation. Today, as well as probably much earlier, various fishes of the
subclass Actinopterygia—such as catfish, gobies (especially mudskippers)—and blennies (especially rockskippers)—cross the strand temporarily. Only the amphibians and their descendants have attained
full and permanent success on land, but except in some specialized
groups the amphibians have maintained partial dependency on
water.

An immense amount has been written about the origin of the
amphibians, covering all aspects, including their ancestry and possible patterns and motivations for passage onto land (reproduction,
protection, feeding, migration to other aquatic habitats in a patchy
environment). But these interpretations have been based largely on
amphibians from the post-Devonian, for which the record is more
suitable as a basis for such speculation. The Devonian contribution
comes less from the amphibians than from rhipidistian fishes of
the order Crossopterygia and dipnoan fishes—variously considered
possible progenitors of the amphibians; these taxa provide some
clues as to possible environments involved in transitions from water
to land.

The earliest known tetrapod or near-tetrapod fossils are from the
late Frasnian of Scotland (Ahlberg 1991). The taxonomic assignment
of these fossils is unclear. The earliest clear evidences of amphibians
include well-preserved materials from Greenland (Jarvik 1952; Coates
and Clack 1990, 1991), meager remains from Australia (Campbell and
Bell 1977), footprints from South America (Leonardi 1982, 1983), and
a fairly well-preserved specimen from Russia (Lebedev 1984). All are
from the Famennian age at the end of the Devonian (Young 1990;
Lombard and Sumida 1992). The Laurasian and Gondwanan specimens were separated by wide expanses of ocean waters and land. A
long and unknown history of the amphibians must have preceded

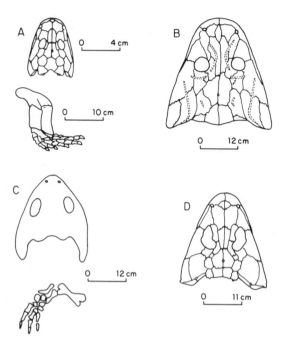

FIGURE   5.6   Early amphibians. (Not drawn to scale.) A: skull and fore-
limb of *Ichthyostega*, late Devonian. *After Jarvik 1952*. B: *Greererpeton*, early
Carboniferous. *After Romer 1960*. C: *Loxoma*, early Carboniferous. *After
Beaumont 1977*. D: unnamed early Carboniferous rhacitomous amphibian. *Redrawn from photograph in Wood et al. 1985*.

the particular taxa represented by these specimens. The specimen
from Russia has been referred to the proterogyrinid order of Anthracosauria (Batrachosauria), well known from the Visean epoch of the
early Carboniferous. The Greenland and Australian specimens appear to be members of the order Ichthyostegalia. Thus, extensive
taxonomic divergence had already taken place, reinforcing the idea
of a remote common ancestor. The Temnospondyli order, which
underwent a major radiation during the late Carboniferous and Permian and survived into the late Mesozoic, have not as yet been found
in the Devonian.

   The amphibians from Greenland and Russia (figure 5.6) were full-
fledged tetrapods, fully capable of walking on land but likely primarily aquatic. *Ichthyostega* measured about a meter in length, *Acantho-*

stega and *Tulerpeton* about half that. The jaw from Australia (*Metaxyg-nathus*) is about twelve centimeters long. Also from Australia are footprints in beds somewhat older than those from which this specimen came (Warren and Wakefield 1972; Warren et al. 1986). The overall length of the animals responsible for these prints has been estimated at a half meter.

These early amphibians were carnivorous and most likely primarily piscivorous. They all were moderately large. They had at least partial metamorphoses, and different groups had variable (4–8) numbers of digits on their limbs (Coates and Clack 1990, 1991). A full transition to tetrapods clearly had taken place by the time that the first fossil amphibians are recorded, and no tangible evidence of transitional stages has yet been found in earlier rocks. No amphibious rhipidistians have been identified.

5.3.2 Theories of Tetrapod Origins

Although the fossil evidence of tetrapod origins is meager, so much has been written on the subject that it cannot be ignored, although it contributes only tangentially to the main thrust of the problem of transition to land. The speculations and interpretations have come primarily from three sources: (1) knowledge of the two groups of Devonian osteichthyans from which the amphibians may have arisen; (2) the fossil amphibians of the Carboniferous and Permian; and (3) the biological (including molecular biological) and morphological information from living amphibians, dipnoans, coelacanths, and amphibious fishes. Recent discussions of these issues include those of Bemis et al. (1987), Panchen and Smithson (1987), Benton (1988), Foreman et al. (1989), Wake and Roth (1989), Schultze and Trueb (1991), Musick et al. (1991), and Marshall and Schultze (1992).

Both crossopterygian and dipnoan fishes have good Devonian records. For many years the consensus has been that the source of amphibians lay among one or more groups of the loose polyphyletic assemblage known as Rhipidistia, within the order Crossopterygia. Very early, however, and finding acceptance from time to time, has been support for the alternative concept that tetrapods arose from the dipnoans. Rosen et al. (1981) presented a thorough study documenting the case for dipnoan affinities, concluding that the tetrapods and dipnoans are sister groups, with the rhipidistians (actinistians) derived separately. The conclusions were based on detailed anatomical studies woven into a cladistic analysis. Rosen and colleagues argue that the long-supposed relationships of rhipidistians and tetrapods are based on general resemblances, fostered by a "search for ancestors" (which the proponents deplore) and expressed in plesiomorphic characters. This point of view has received mixed reactions and was vigorously attacked by Schultze (1981), Jarvik (1981), Panchen and Smithson (1987), and Marshall and Schultze (1992).

One focus of these interpretations that relates to a search for the pathways of land invasions by tetrapods is the aquatic environment from which they stemmed. Remains of the earliest amphibians come from freshwater deposits. Rhipidistians and dipnoans in post-Devonian rocks are largely from freshwater, and occasionally brackish beds, except for transported remains and some apparent marine amphibians from the Triassic. In the Devonian, however, the early remains of dipnoans have come from marine rocks, whereas those of rhipidistians are primarily from freshwater, although some of the species are found in rocks formed under a variety of depositional conditions, suggesting they may have been anadromous. If the amphibians originated early in the Devonian, as their full development by the late Devonian suggests, an origin from Dipnoi might indicate a marine source, whereas one from Rhipidistia could suggest a freshwater source. Most of the scenarios of origin have been based on the second interpretation. Bray (1985) provides an interestingly different scenario for the origin of tetrapods based on a marine rhipidistian ancestor.

It has been generally assumed that the predecessors of the Amphibia already breathed air by use of lungs. This may well have been the case, but direct evidence is nonexistent. A wide variety of fishes, of course, can survive and be active for some time out of water. Detailed investigations (chapter 9) have demonstrated that many amphibious marine teleosts respire efficiently in the atmosphere under moist conditions completely without lungs. If this were true for the lineage(s) that led to amphibians, which is at least possible, the evolutionary stage at which lungs appeared did not necessarily precede the origin of amphibians.

A major phylogenetic question that has bearing on the patterns of transition to land in tetrapods is whether the amphibians are a monophyletic or polyphyletic group. Once again, a monophyletic interpretation has been generally made (Hillis 1991), although some (Holmgren 1939; Jarvik 1955) have suggested various patterns of polyphyletic origins. Much of the evidence for either interpretation comes not from the Devonian amphibians but from those of more recent times. This issue will thus be considered in the next chapter, and we present our own analysis and synthesis of the data in the final chapter.

Whatever the precise ancestry, it is probable that the major features of the anatomical arrangements and the structures of tetrapod limbs evolved in completely aquatic fishes—either freshwater or marine. Thus these features should be viewed as examples of preadaptations that helped facilitate, if not actually enable, the water-land transition. A cladistic analysis of limb morphologies in sarcopterygian fishes (the subclass that includes both rhipidistians and dipnoans) indicates that the closest fossil relatives of the tetrapods were the aquatic rhipidistian crossopterygians of the family Osteolepidae (Edwards 1989). Additional support for this view derives from anatomical studies on newly discovered fossils of *Acanthostega gunnari*, from the late Devonian of Greenland. Analyses of limb structures in these animals indicate that limbs with digits probably first evolved for use in water, rather than for walking on land. The numbers of digits on these limbs varied from four to eight (Coates and Clack 1990, 1991).

How, when, and for what reasons were the metamorphoses carried out by many living amphibians first evolved? The answers are all unknown. The Devonian fossil record is largely silent on the questions of whether or not, or to what extent, the first amphibians underwent metamorphoses as part of their life cycles. An assumption made by some workers is that metamorphosis probably did occur in at least some forms, but it was a substantially less profound set of processes than it is in modern amphibians. Support for this position derives from additional work on *Acanthostega gunnari*. The recently found fossils of this genus of ichthyostegalian contain a fishlike branchial skeleton, fishlike internal gills, and an open opercular chamber suitable for use in aquatic respiration (Coates and Clack 1991). Evidence for metamorphoses like those of modern amphibians first appears in the fossil record of the Triassic (Duellman and Trueb 1986:171).

To the extent that the earliest tetrapods used lungs for aerial respiration, their sprawling postures probably limited their abilities to move rapidly for anything more than short distances. Carrier (1987) presents a variety of evidence, based on studies of both fossils and living crocodilians and lizards, that the earliest tetrapods probably could not simultaneously breathe and actively walk or swim. If this was so, these forms were probably sit-and-wait predators. The extent to which they may have used cutaneous and/or buccopharyngeal respiration is unknown. They probably also had limited powers of geographic dispersal and were almost certainly not migratory to any substantial degree. Nothing is known about their larval or juvenile stages. The anatomical linkages between the locomotor and respiratory systems of the early tetrapods probably also placed severe constraints on the evolutionary sequences of events that led to the development of breathing with lungs.

To close this section we point out again that no fossil fishes have been identified as amphibious. We also know of no fossils that relate directly to the evolutionary origins of any of the living amphibious fishes. Thus there is no direct evidence available bearing on the questions of whether or not there may have been partial or otherwise failed additional vertebrate water-land transitions or of how long the extant lineages of amphibious fishes have been amphibious. The underlying problem here may well be the basic morphological similarities of living amphibious forms to their close aquatic relatives. There is no reliable way to tell them apart.

5.4 Summary

The sediments of the late Silurian have yielded incontrovertible evidence of terrestrial metazoans, including a trigonotarbid, a true millipede, and several genera of centipedes. During the Devonian, exploitation of continental environments by both plants and metazoans expanded rapidly, a precursor to the extensive occupancy of land that would occur during the Carboniferous. The Old Red Sandstone, which stretches across the British Isles, Scandinavia, Greenland, and northeastern North America, covers a time span from late Silurian into the Devonian and has produced the most instructive sequences of metazoans and plants. Three sites, the Rhynie Chert of Scotland, the Alken site of Germany, and the Gilboa site of New York, have produced the bulk of known invertebrates. Arthropod metazoans predominate and are associated with early terrestrial vascular plants. Each of the three sites was formed in a distinctive environment, and together they have produced a diverse sample of Devonian terrestrial metazoa, including trigonotarbids, arthropleuran myriapods, mites, scorpions (?aquatic), and a collembolan hexapod. The animals were predominantly carnivorous and, to a lesser degree, detritivorous, and their food chain presumably was based in algae.

No fossils are know that relate directly to the vertebrate transitions to land. No amphibious rhipidistian crossopterygians have been identified. Recent fossil finds indicate that the tetrapod limb probably first evolved in aquatic environments. These finds also make it likely that pentadactyly was only one of several early conditions relating to numbers of digits on early limbs. The early transitional forms had internal gills and may or may not have had lungs.

The earliest evidence of amphibians comes from late Devonian sites in Greenland, South America, Russia, and Australia. The Greenland amphibians, primarily ichthyostegids, are by far the most complete. A jaw and some footprints are known from Australia (tentatively assigned to the ichthyostegids). A single partial skeleton, possibly a batrachosaur, comes from Russia and footprints from South America. The most likely sources of these amphibians lay among the rhipidistian crossopterygians. The geographic distribution and morphological diversity of the fragmentary remains has posed problems and has led to controversy as to whether amphibians are monophyletic or polyphyletic and whether they arose from freshwater or marine fishes. The scantiness of the evidence has also made it difficult to reconstruct patterns of dispersal and genetic relatedness among the earliest terrestrial tetrapods and their progenitors. Most of the answers have thus been sought by studies of the much better known amphibians in the Carboniferous and Permian Periods. However, these more recent forms were temporally far removed from the initial stages of the vertebrate transitions to land.

6

Fossil Evidence of Metazoan Transitions: Permo-Carboniferous and Later

EVERETT C. OLSON

6.1 The Environmental Setting

6.2 Arthropods
    6.2.1 Chelicerates
    6.2.2 Crustacea
    6.2.3 Uniramians

6.3 Molluscs

6.4 Annelids: Leeches

6.5 Chordates: Tetrapods
    6.5.1 Amphibians (Anamniotes)
    6.5.2 Reptiles

6.6 Summary

Only the Pridoli site and the three sites from the Devonian as described in chapter 5 have yielded significant assemblages of terrestrial arthropods. Each of these sites tapped different depositional environments, suggesting that land invasion by the arthropods was well under way by the middle Devonian. Very few terrestrial sites have been found in rocks of the early Carboniferous; these mostly add little to the evidence of the extent of land occupancy. One early Carboniferous site, however, has added importantly to our knowledge. It is located near Bathgate, West Lothian in Scotland (Wood et al. 1985; Smithson 1989) and has produced a unique assemblage of animals. The invertebrates include myriapods, scorpions, and a giant eurypterid—all belonging to groups known in the Devonian—plus the earliest known harvestman (Opiliones). In addition, a well-preserved amphibian from the order Temnospondyli and an articulated skull and skeleton of a small animal tentatively identified as a reptile have been found. Beyond extending the temporal reach of the reptiles, the Bathgate site is extremely important in studies of early ecological associations of predominantly terrestrial and semiterrestrial animals.

In contrast, widespread and taxonomically diverse assemblages of terrestrial animals have been found in the late Carboniferous and into the Permian. These Permo-Carboniferous assemblages offer evidence of a major adaptive radiation—especially among arachnids, flying insects, amphibians, and reptiles. During the Permo-Carboniferous, in contrast to earlier times, radiations of existing terrestrial lineages became the dominant feature of evolution; invasions of the continents by new types of organisms arising out of aquatic lineages may have come to a halt. Rather, the terrestrial radiation was matched by extensive radiations of freshwater animals, especially crustaceans and fishes, and amphibians. Amphibians were moderately diversified by the time extensive records occur in the late Carboniferous.

A number of additional important events in metazoan transitions to the land occurred much later—in both the Mesozoic and the Cenozoic. These include the origins of many of the modern groups of terrestrial molluscs, of land leeches, of all living groups of amphibious and terrestrial crustaceans, and of all living groups of amphibians. Indeed, while perhaps improbable, it is possible that still other metazoan groups are in process of invading the land even today.

6.1 The Environmental Setting

Vascular plants had expanded to form forests by the late Devonian (chapter 3); they continued terrestrial radiations during the rest of the Paleozoic. Early Carboniferous floras, recorded in deposits formed along sea margins, expanded to form the great, tropical coal swamps that spread across Laurussia during the late Carboniferous. As climates changed during the Permian and into the Triassic, the tropical forests were succeeded by more temperate floras as higher and drier lands were occupied. Changes in vegetation were accompanied by expansions of terrestrial animals in a continuing coevolutionary pattern (Kevan et al. 1975; Little 1983, 1990; Shear and Kukalova-Peck 1990; Behrensmeyer et al. 1992).

Figure 2.8 illustrates the main trends of continental development during the Permo-Carboniferous. Continued continental movements and collisions produced a nearly completed continent of Pangea by the end of the Paleozoic, with the final aggregation occurring during the latest Triassic. The mountains of the fold belts formed by the collision of Laurussia and Gondwana (Hercynian, Appalachian, Ouachita, and Mauritanide mountains) appear to have altered the flow of equatorial easterly winds, leading to replacement of the humid, warm tropical belt by a drier zone and widespread equatorial aridity by the end of the Permian period. Permian nonmarine fossils, in contrast to those of the Carboniferous, come from sediments formed under increasingly temperate circumstances.

Gondwana remained a gigantic land mass, with slowly diminishing polar expanses as it moved northward and rotated, eventually meshing with the margins of Laurussia. During the Carboniferous and very early Permian, extensive continental glaciation prevailed in the southern regions, but this does not appear to have had drastic climatic effects on the equatorial swamp belt. Cyclical rises and falls of sea level, however, were probably related to waxing and waning of glacial cycles in the Antarctic and produced shifting shallow marine and brackish and freshwater habitats associated with the coal swamps.

6.2 Arthropods

As during the Devonian, arthropods were the dominant terrestrial metazoans during much of the Permo-Carboniferous. Arachnids and insects continued their expansions and were accompanied by rapidly differentiating lines of amphibians, reptiles, and land snails—thus greatly increasing the complexity of terrestrial ecosystems.

Chelicerate and uniramian arthropods were highly diversified during the Permo-Carboniferous. The preponderance of fossil remains have come from the coal measures of North America and Europe, with remains of arachnids and pterygote insects predominant. They occur in the coal itself, and in associated shallow marine and brackish deposits. Some of the best preserved remains have come from siderite nodules formed in the clays associated with coals at Mazon Creek in Illinois and similar nodules from Montceau-les-Mines, France. Crustaceans are well represented in deposits of the coal measures, but during the Carboniferous, as today, continental crustaceans were predominantly freshwater animals, and relatively few attained complete emancipation from water. On land their greatest success has been primarily in cryptic environments.

All of the major groups of arthropods that had emerged onto land by the late Carboniferous presumably developed during the poorly known adaptive radiations of the Devonian and early Carboniferous. These radiations probably took place in large part on land, although this is not testable in the fossil record. Only the scorpions give fossil evidence of the stages of transition from marine to terrestrial environments, as seen both in their morphology and the sedimentary environments in which they have been found (Størmer 1976). Intermediates are either unknown or unrecognized in other major groups.

6.2.1 Chelicerates

Of the three classes of Chelicerata listed in table 4.1, only the Arachnida made successful transitions to land. Sixteen orders of terrestrial arachnids are listed (after Petrunkevitch 1953). Three of these are extinct and only two lack a Carboniferous record. Schizomida, absent in the Paleozoic, has been found in Cenozoic Baltic ambers, and Palpigradi has no fossil record. During the Carboniferous, as today, the arachnids were a dominant group of terrestrial arthropod carnivores. Similarly, they and the insects probably became important prey of other small carnivores, as amphibians and reptiles became important elements in the trophic web during the late Carboniferous.

Eurypterids As discussed earlier, eurypterids never evolved fully terrestrial members, at most being confined to intertidal and possibly supratidal environments. On the basis of analyses and interpretation of the structure of the gills, following Wills (1965) and Waterson (1975), Størmer (1976) concluded that the gills in some eurypterids were rather similar to those of some terrestrial isopods and that this, coupled with evidence from tracks, suggests that some eurypterids may have been amphibious during the late Paleozoic.

Xiphosurans Fisher (1979) made an extensive analysis of *Euproops danae*, the best known of the fossil xiphosurans. *Euproops* comes from the Essex fauna of Mazon Creek, Illinois, which includes a variety of well-preserved, soft-bodied animals from shallow water marine beds of the late Carboniferous coal measures. He concluded that this genus, and perhaps other xiphosurans, were "occasionally present and active in a subaerial context." Horseshoe crabs today venture into intertidal areas for short periods and, as suggested by Fisher, at least some of their Paleozoic relatives may have made even more extensive use of semiterrestrial environments. The xiphosuran pathway to land was clearly from marine waters, without a freshwater phase. The xiphosurans are an ancient group, dating back to the late Silurian or early Devonian. During this long history forays onto land were likely characteristic. Both the morphological and physiological aspects of these organisms appear to have been unusually stereotyped throughout their history, and none of them appears to have maintained more than temporary emancipation from marine life. Like eurypterids, xiphosurans apparently failed to evolve any fully terrestrial members because of constraints imposed by structural and physiological characteristics, which, in turn, failed to evolve in directions that would have made full terrestriality possible.

6.2.2 Crustacea

Crustaceans are predominantly marine arthropods, although freshwater and terrestrial representatives have evolved in five of the eight classes. Only the Cephalocarida, Mystacocarida, and Cirripedia are strictly marine. Among the other five remaining classes, terrestrial representatives are primarily cryptic, as shown in table 4.1. Noncryptic terrestrials occur among the malacostracan orders: Isopoda, Amphipoda, and Decapoda. The marine record of Crustacea begins in the Cambrian; the freshwater record in the mid-Paleozoic. Cryptic and epifaunal terrestrial representatives have a scanty record, in part because of the difficulty in distinguishing from fossil evidence the aquatic from the amphibious species and also because of the rather special circumstance in which epifaunal types occur.

Invasions of the continents certainly took place independently in each of the five classes, and, from the evidence of extant representatives, multiple invasions likely occurred in most of the subgroups with terrestrial forms. The anatomical structures and presumed phylogenies of the malacostracans, as well as their environments, indicate multiple pathways to land from both marine and freshwater

environments. Once again, however, the scarcity of fossils and difficulties of environmental interpretations provide little information
about the precise histories of these transitions.

The result is that the times of origins of the continental groups
remain uncertain, although the occurrences probably covered a wide
span of geological time.

Branchiopoda Branchiopods (fairy shrimp, brine shrimp, and water
fleas) today occur mainly in fresh and highly saline waters. Best
known are the fossil conchostracans, which nevertheless have only a
spotty record from the Devonian to the Recent. They are most often
found in fine-grained sediments in association with amphibians and
freshwater fishes. Cryptic phases are present among extant branchiopods, but there is no way, from the evidence available in the fossil
record, that this type of existence in the past can be verified except
by associations with other animals—and these associations may be
dubious because of the chances of depositional mixing.

Ostracoda Like the branchiopods, marine and freshwater ostracods
have a long fossil history. Today cryptic species exist and these
appear to be closely related to freshwater species, suggesting freshwater as the probable ancestral habitat. Freshwater ostracods, like
the branchiopods, occur in association with freshwater animals of
various other taxa, both vertebrates and invertebrates. No unequivocal evidence of cryptic terrestrials is found in the fossil record.

Copepoda Although copepods are dominantly marine and freshwater inhabitants, a number of species have abundant cryptic, terrestrial
representatives today. They live in leaf molds and require water for
an active existence. Some species form cysts and thus survive periods
of desiccation. No fossil records of such occurrences are known.

Malacostracans have been more successful on land than have the
other classes of crustaceans. Nevertheless, only relatively few groups
have adapted to a completely terrestrial existence. Two orders of the
Peracarida—Isopoda and Amphipoda—have had great success on
land, and members of several lines of decapods have attained partial
or complete terrestriality. (Adaptations of these groups are discussed
in chapter 8.)

Isopoda On the basis of both numbers and diversity of environments the oniscoid isopods (sow bugs and pill bugs) are the most successful of terrestrial invertebrates today; they are rivaled only by arachnids and insects. The members of the other isopod subgroups are predominantly marine. Morphological and physiological differences within the suborder Oniscoidea reflect adaptations to various aspects of terrestrial life; the existence of twenty-one families (Little 1983) illustrates the extensive adaptive radiation oniscoids have undergone. A few living species are amphibious.

The fossil record of oniscoids is scant and, as Edney (1968) and Schram (1986) have indicated, gives little information on the origins of the various families. The record of terrestrial oniscoids goes back only to the Eocene. The order as a whole goes back to the Triassic, by which time all of the major marine groups were present. The terrestrial oniscoids could have developed at that time or at any later time during the Mesozoic. Vandel (1943, 1965) concluded that the major groups arose independently from different marine stocks, but the fossil record gives no reliable information on when these origins may have taken place or the pathways followed. What is known and the conclusions that have been reached have come almost entirely from extant species.

Amphipoda A single family, Talitridae, with some two hundred known species, includes cryptic and semiterrestrial members, but it is predominantly an aquatic group. Some semiterrestrial talitrid species (sand and beach hoppers) inhabit littoral marine environments, but most occur in leaf litter in forests. None is adapted to strictly terrestrial life. The littoral species suggest origin from marine sources, but the very poor fossil record, mainly Neogene, gives no additional evidence of sources or times of origin.

Decapoda Twelve decapod families with terrestrial or semiterrestrial members are listed by Bliss (1968). Schram (1986) places these animals in eight families. The suborder Astacidea (crayfish) nearly bridge the gap between water and land; in addition to forays onto land, some crayfish survive dry periods in burrows that tap the water table (Bliss 1979). None are completely terrestrial. Constraints have been imposed by apparently unalterable structures and physiology.

The brachyuran family Gecarcinidae comprises the terrestrial land crabs. The extant *Gecarcinus lateralis* has solved many of the problems of terrestrial existence. Most genera occur in tropical and semitropical islands. *Birgus latro*, the coconut crab, is one of the most highly terrestrial of all living decapods. Its reproduction, however, maintains a tie to water. Many members of the other two brachyuran families listed in table 4.1 are amphibious.

Among the hermit crabs (anomuran family Coenobitidae) there is a habitat and structural gradation from entirely marine to semiterrestrial animals, in which capacities for air breathing have developed. The main point of interest attached to this gradation is that it occurs within a limited familial lineage and not in the course of some major morphological shift.

Decapods have a fossil record that extends from the late Devonian to the present. The record is fragmentary both stratigraphically and geographically. No distinctly terrestrial fossil decapods have been recognized; in view of their limited structural differences and their habitats this is to be expected. Two major morphological types—Natantia (marine) and Reptantia (freshwater and terrestrial)—are distinguished by absence of "walking-legs" in the former and their presence in the latter. The brackish to freshwater species, and those in shallow marine and strand line deposits, belong to the "walking" group of Reptantia.

Both shallow water and nonmarine decapods are known from the Carboniferous. Their diversity in form and habitat (brackish to probable freshwater) suggests that the base for land invasion was well established and the transitions may have taken place prior to this time (Briggs and Clarkson 1989). Burrows in freshwater Permo-Carboniferous sediments, in which there is evidence of occasional dry periods, may have been made by semiterrestrial decapods, but remains of the animals that may have been responsible for them have not been identified. The various pathways from water to land suggested by extant decapods can neither be confirmed nor denied by the fossil representatives, but the variety of structures and habitats that they display makes the likelihood of multiple invasions high.

The earliest known fossils identified as belonging to one of the extant amphibious or semiterrestrial decapod families dates from the Jurassic. To the limited extent that fossils are known from the other families, they have Cenozoic (Eocene, Pliocene) ages (Schram 1986).

Other indirect lines of evidence, primarily zoogeographic, that might provide some information relating to places and times of origin of the living amphibious to terrestrial groups are notably confusing for all groups (Schram 1986).

6.2.3 Uniramians

The three uniramian subphyla—Onychophora, Myriapoda, and Hexapoda—are present in the late Carboniferous. The class Arthropleurida, a prominent group at that time, is also sometimes considered as forming a separate subphylum of Uniramia. For convenience here, the arthropleuridans will be considered as myriapods, with recognition that their position is a matter of judgment (Rolfe 1969; Manton 1977).

Records from earlier deposits show that uniramian invasions of land (by collembolans and millipedes) were underway as early as the beginning of the Devonian, but only scant evidence records their presence from the middle Devonian through the late Carboniferous. The most prolific sites of the late Carboniferous are Mazon Creek in Illinois and Montceau-les-Mines in Autun, France. In both these sites preservation is excellent. Uniramian fossils are also abundant in Joggins, Nova Scotia, where fragments of many species of arachnids occur in and around fossilized, upright sigillarian tree trunks. There can be no doubt that the major land invasions of uniramians had been well established by the late Carboniferous and that radiations thereafter ensued primarily from exploitation of a range of terrestrial environments (Behrensmeyer et al. 1992). The few living groups not present in the fossil record or present only in the Cenozoic, may, of course, have come onto land later, but a more probable explanation of their absence is the accidents of preservation and discovery.

Onychophora The onychophorans are more or less prototypical lobopods. The first marine record of this structural group, as noted in chapter 4, is the genus *Aysheaia* from the middle Cambrian, although a possible onychophoran, *Xenusion*, is somewhat earlier (Dzik and Krumbiegel 1989). Following an extended gap in time the next representatives occur in Carboniferous deposits, from Montceau-les-Mines and Mazon Creek. The specimen from Montceau-les-Mines clearly is an onychophoran and has been confidently assigned on the basis of

well preserved antennae (Rolfe et al. 1982). The Mazon Creek specimens were described by Thompson and Jones (1980) and tentatively
assigned to Onychophora. The resemblance of the two finds supports the tentative assignment of the Mazon Creek specimens to this
group. Both probably were terrestrial, with habits much like those
of the living *Peripatus*, but they occur in a mixed terrestrial-marine
environment that does not itself testify to terrestrial habitation.

No intermediates occur between the late Carboniferous and modern onychophorans. This emphasizes once again the problems in
interpreting the histories of soft-bodied organisms and speculating
on the pathways their lineages took in invading the land.

Myriapoda   Three classes of myriapods—Chilopoda, Diplopoda, and
Arthropleurida—are well represented in the late Carboniferous.
Pauropoda has no fossil record and Symphyla appears only in Pa-
leogene ambers. When each came into existence is not known, all
but Arthropleurida have survived to the present.

Centipedes (Chilopoda) first appear in the record in the middle
Devonian, and they have an excellent record in the late Carboniferous of Mazon Creek (Mundel 1979). The two known genera, *Mazonoscolopendra* and *Latzelia*, have been placed in separate subclasses,
Epimorpha and Anamorpha respectively. They are distinctly modern
in character; evolution in the group since the late Carboniferous
has been extremely conservative. The Carboniferous centipedes were
fully terrestrial and their diversity suggests a fairly extended evolutionary history.

Millipedes (Diplopoda) were highly diverse during the late Carboniferous. A wide variety of probable species have been described
from the Joggins region of Nova Scotia. Some resemble the burrowers and others the debris-splitters of present day millipedes, but still
others are more specialized than any living forms, being set with
large, defensive spines. Some, like the related arthropleurids, attained lengths up to a third of a meter. Diplopodan myriapods, along
with the relatively small arthropleurids, occur at Montceau-les-Mines
(Rolfe et al. 1982; Heyler 1985–86).

Millipedes, centipedes, and arthropleurids are known from the
early Devonian. The intervening record is poor, but the Serpukhovian Carboniferous has yielded evidence of their presence in the form
of trackways (Briggs et al. 1979). Again, however, as is the case for

all uniramians, no evidence of transitions from aquatic to terrestrial habitats has been found, suggesting, as Manton (1977) indicated, that they originated on land from unpreserved, soft-bodied ancestors of an intermediate type. The putative myriapod from the Cambrian, as well as morphological evidence, argues to the contrary. The conservative millipede types have persisted to today, but the apparently more specialized forms, the arthropleurids, appear to have become extinct by the late Paleozoic.

Hexapoda The record of apterygote hexapods (wingless insects), after their early Devonian occurrences, is scant. Specimens from the Gzelian Carboniferous and the Permian form the basis for the order Monura of Sharov (1957). The genus to which some of the specimens of monurans have been referred was first described as a crustacean and its position has been a matter of dispute. Carpenter (1977), in his short review of fossil hexapods, suggested that Monura resembled extant archaeognathans, and it has been suggested that the mid-Devonian apterygote from the Gilboa site may belong to this group as well. Specimens from the Eocene and Miocene of Europe and North America nearly complete the roster of fossil remains of this group. In addition, thysanurans and archaeognathans occur in amber of Cretaceous, Eocene, Miocene, and Oligocene age (Poinar 1992). The structure and associations of the fossil apterygotes indicate that throughout their history they formed a strictly terrestrial group, primarily cryptic. No aquatic forerunners have been identified.

A rich and diverse assemblage of pterygote hexapods (winged insects), both Palaeoptera and Neoptera, was present in the late Carboniferous, with both of these major subdivisions represented by several orders. An extreme proliferation in naming led to the designation of some fifty orders. This gives a false impression of the extent of the Carboniferous radiation, but even the taxonomic reappraisal by Carpenter (1977) accepts a diversity that demonstrates very extensive adaptive radiation. Carpenter conservatively recognized eleven orders, seven of which are now extinct. Kukalova-Peck (1991) suggested that a classification of winged insects into thirteen to fifteen orders was more realistic. Various considerations of the systematics of fossil insects have since modified Carpenter's views but have not materially altered the general picture of the extensive radiations (Rohdendorf and Rasnitsyn 1980; Hennig 1981; Wootton 1981; Boudreaux 1987; Kristensen 1989). The end of the Paleozoic appears to have witnessed the termination of the first great wave of pterygote expansion, followed by a second wave beginning in the mid-Mesozoic and continuing to the present. Both radiations were primarily terrestrial as far as adult phases of the various groups are concerned.

The first of the winged insects are found in rocks of Serpukhovian age, near the boundary that distinguishes the late from the early Carboniferous (Kukalova 1964; Wootton 1981). An explosive evolution followed—perhaps actually beginning as early as the Devonian—during which insects emerged from their cryptozoic associations to become an important part of the developing epifaunal coevolution of plants and animals in the terrestrial realm. Their radiations were accompanied by, and perhaps to some extent preceded, similar expansions of the arachnids. Both insects and arachnids were intimately related to developing vertebrate penetrations of the land and rapid diversification of terrestrial floras.

The insect expansion, well documented beginning in the late Carboniferous, continued through the Permian. During this time the number of orders of insects approximately doubled, and of these seventeen are now extinct. Both the Paleozoic and the Mesozoic radiations were basically terrestrial, finding their beginnings in the apterygotes. Some reinvasions of water took place (Messner 1988) and, of course, larval development in water is characteristic of many orders. The initial phases of land occupancy are not known but seem to have involved land occupancy by small, cryptic apterygotes, followed by a spread to more strictly terrestrial environments, and continued adaptive occupancy of new physical and biological niches as evolutionary changes in other groups of organisms provided new avenues for proliferation.

6.3 Molluscs

Of the immense phylum Mollusca, only the class Gastropoda has evolved into terrestrial animals. Bivalvia, of course, are common inhabitants of freshwater and have been from at least as early as the

Carboniferous or perhaps early Devonian. No terrestrial bivalves have evolved even though some have developed air-breathing capacities. Immobility appears to have been the limiting factor.

Gastropods have been successful in a wide range of environments both in the water and on land. (Chapter 7 describes the adaptations of living forms.) Terrestrial snails have appeared in the fossil record at several different times, with major deployments recorded in the late Carboniferous and again in the Jurassic-Cretaceous. Although their evolution has been conservative on land, it is probable that major terrestrial groups, perhaps at the superfamily level, have evolved independently from aquatic ancestors. On the basis of structure, habits, and relationships, it may be inferred that several different pathways to land life were involved (Graham 1985). No fossil molluscs are known that are identifiable as amphibious or semiterrestrial. The principal suborders that have terrestrial representatives are listed in table 4.1.

Solem and Yochelson (1979), Solem (1979, 1985) and Little (1983, 1990) have given extensive accounts of the terrestrial snails, with emphasis in the former papers on the fossil record of the Paleozoic. The first known fossils of land snails, prosobranchs of the genus *Dawsonella*, occur in two rather unusual circumstances in the late Carboniferous—in freshwater limestones and in the upright fossil tree stumps at Joggins, Nova Scotia. *Dawsonella* has been tentatively assigned to the archaeogastropodan neritacean family Helicinidae. Fossils of marine neritaceans are present in mid-Devonian and early Carboniferous sediments. The second oldest terrestrial prosobranch group preserved in the fossil record is the family Pomatiasidae, which appears in rocks of Cretaceous age (about 100 mya). The third oldest family, Cyclophoridae first appears in rocks of Tertiary age (about 50 mya).

Four families of pulmonate Stylommatophora (Tornatellinidae, Pupidae, Enidae, Discidae) are present in late Carboniferous deposits, showing a major diversification of terrestrial forms by this time. The four families of pulmonate Stylommatophora, as well as the prosobranch Helicinidae, are still in existence today.

The fact that the first known fossils for each of these groups are similar to extant terrestrial families may imply any or all of several things: (1) The actual evolutionary transitions to land in these lineages occurred substantially earlier than their first known appearances in the record. (2) The earliest appearances were actually of amphibious, not completely terrestrial forms. (3) The morphological stability of the shell types over huge stretches of time may mean that the soft anatomy of these groups has been similarly stable, and that their present functional properties and ecological relationships represent comparable continuity.

A long gap in the fossil record of land snails occurs from the early Permian up to the late Jurassic. The pulmonate order Basommatophora, generally considered more primitive than the order Stylommatophora, first appears in the late Jurassic, indicating a very fragmentary fossil record, some misclassification of fossils, or a misinterpretation of phylogenetic relationships. The first explanation seems more probable. The Jurassic appears to mark the start of the extensive radiation which has led to the great variety of land snails today. By the end of the Paleozoic the basis for much of the radiation, excepting that of the Basommatophora, was established. In view of the long gap in the fossil record, expansion may have begun at any time after the beginning of the Mesozoic era.

6.4 Annelids: Leeches

No fossils have been unequivocally identified as leeches (class Hirudinea). What is known of the adaptations for life out of water possessed by living land leeches is summarized in chapter 8.

Sawyer (1986) analyzes what he considers the relevant indirect evidence of the origin of the class Hirudinea—including the fossil histories of animal groups with which living leeches have unusually specific relationships, leech zoogeography, and the distributions of leech species considered relicts. His general conclusion is that the class Hirudinea probably first arose in the early Mesozoic. The zoogeographic data, much of it specifically based on the distributions of specialized groups of land leeches, indicates that the two orders of leeches are probably also of comparable antiquity. There is no way of determining whether or not the ability to live fully terrestrially is similarly ancient (thus terrestriality could be considered to have a nearly monophyletic origin), or if the wide geographic dispersion of terrestriality in living leeches represents multiple independent origins of that life history feature (polyphyly).

Whatever may have been the origins of terrestrial capabilities in the different groups, the three families of arhynchobdellid land leeches are each widely distributed. The Cylicobdellidae are known from South America, east Asia, and many of the Pacific islands, including the Hawaiian islands. The Americobdellidae are South American. The Haemadipsidae are by far the most widely distributed, occurring in southern South America, southern Africa, south and southeast Asia, Australasia, some southern European countries, and various islands in the western Indian Ocean, including Madagascar. The populations on isolated oceanic islands are considered to have derived from leeches carried by birds.

6.5 Chordates: Tetrapods

We now consider the later stages in vertebrate transitions to land. The conventional view of the origin of terrestrial tetrapods includes the following interpretations:

1. The initial strictly terrestrial tetrapods were amniote reptiles.
2. The amniotes arose from anamniotes (amphibians), which were not fully terrestrial.
3. The amniote tetrapods are monophyletic.
4. The transition from semiterrestrial to terrestrial existence took place no later than the very early Carboniferous.

These interpretations provide a plausible framework for a scenario of the final stages of the invasion of land by tetrapods. Each, however, is open to question, based both on new information and the systematic philosophy under which it is cast. These questions are raised and examined in the following sections. They are also treated in much more detail, mostly within a cladistic mode, in Schultze and Trueb 1991.

6.5.1 Amphibians (Anamniotes)

Following the late Devonian records of amphibians in Greenland, Australia, and Russia, no further evidence of tetrapods has been found in rocks deposited prior to those of the Upper Visean of the early Carboniferous. Upper Visean sites in West Virginia, Iowa, Nova Scotia, and Scotland have produced an interesting but sparse record of amphibians and the remains of the first purported amniote (reptile) (Bolt et al. 1988; Smithson 1989; Lombard and Sumida 1992). The succeeding Serpukhovian and Bashkirian ages, which bridge the boundary between the early and late Carboniferous, have produced more extensive remains. All major groups of extinct amphibians and firmly identified reptiles occur in the Moscovian of the late Carboniferous. The evolutionary radiations recorded in coal measure depositions continued during the Permian, with new and more highly terrestrial representatives of both the reptiles and amphibians evolving as drier, more temperate climates came to be the principal habitats of terrestrial animals.

Members of the class Amphibia are usually partially aquatic and partially terrestrial and are characterized by anamniote eggs, most of which require an aqueous medium for maturation. Various species of extant salamanders and anurans have overcome the strict tie to aquatic habitats in a variety of ways and have become, for all intents and purposes, terrestrial animals. Some of the extinct amphibians, particularly those of the Permian, developed skeletal features that indicate that they too were primarily terrestrial. The evidence of anamniote eggs or analogous structures among the extinct amphibians is conjectural, but in most cases the structure of skeletons, and the habitats and associations of the animals, indicate the validity of such speculation. There is no clear-cut evidence of phylogenetic relationships between the ancient terrestrial amphibia and living forms (Lombard and Sumida 1992).

Figure 6.1 shows the stratigraphic distributions of the major lines of amphibians as classified in table 4.1. The earliest known amphibians of the early Carboniferous are from the Visean age. Although precise time correlations between the Scottish and North American occurrences have not been established, whatever differences exist are insignificant in view of the scantiness of the record. Sparse as it is, this record shows a high level of diversity, even at the beginning of the fossil appearances. The earliest known Lissamphibia are from the Triassic.

The Devonian ichthyostegalians show that capacities for terrestrial locomotion were fully developed by that time. These animals, as well as the later Visean forms, must have had long histories of which we have no knowledge (Milner et al. 1986). This leaves the vertebrate pathways to amphibious and terrestrial life conjectural as far as the fossil record is concerned. It has been mainly from studies of later

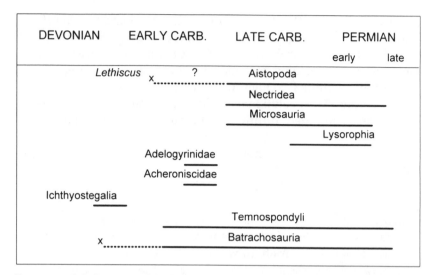

FIGURE   6.1 Stratigraphic distribution of main groups of Paleozoic amphibians. (*Lethiscus* is an early Aistopod.) *See also Lombard and Sumida 1992.*

evidence, including modern amphibians and amphibious fishes, that various concepts of the origin of tetrapods and pathways onto land have been developed.

The nature of the earliest amphibians, however, is of interest in this matter if only because they introduce some constraints on the interpretations. First, of course, is the fact that there was extensive diversity and at least some terrestriality in the very early record. Second, the Paleozoic record shows that the classification of amphibians in table 4.1, which is more or less generally used, is a gross oversimplification of the real situation. Without elaboration of the intricate aspects of the problem, it can be noted that the two major subclasses of amphibians—the Labyrinthodontia and the Lepospondyli—are each, quite surely, polyphyletic. This has long been recognized for the latter and is becoming more and more evident for the former. The three to five clades of "lepospondyls," depending on how associations are made, have no common base either in phylogeny or morphology except for spool-shaped vertebral centra. Striking, but perhaps insignificant temporally, is the fact that the earliest known lepospondyl fossil, *Lethiscus* (of Visean age), was secondarily

limbless and aquatic. It is tentatively assigned to the order Aistopoda, which has a moderately long subsequent history (Wellstead
1982). Unless a polyphyletic origin of the Amphibia is granted, which
has been advocated in several guises, a long history, perhaps as
much as tens of millions of years, would seem to be required to
produce this limbless stock from assumedly limbed ancestors. Pertinent to this is the fact that one of the principal bases for assuming
monophyly of Amphibia lies in the close resemblances of *all* tetrapod
limbs, except, of course, in cases where they have been modified for
special life styles, including near or total loss among some amphibians in addition to Aistopoda and in various reptiles.

The apparent diversity of the class Amphibia shown in table 4.1 is
somewhat lessened if the adelospondyls, an equivocal group, and
the lysorophoids are placed with the microsaurs, as is sometimes
done. This is probably, however, an unjustified "lumping" and is a
matter of convenience based on the inability to establish any reliable
major relationships between the constituent members.

The other major subclass of Amphibia, the Labyrinthodontia, has
generally been considered to be a monophyletic group, with its relationships to the lepospondyls very uncertain. Panchen (1980) and
Panchen and Smithson (1987) have seriously questioned this interpretation, as well as the monophyly of the Temnospondyli and some
of its subordinate groups. The Batrachosauria (Anthracosauria) do,
under Panchen's interpretation, represent a monophyletic assemblage.

Regardless of how these relationships are drawn in detail, they indicate
extensive early diversity of amphibious tetrapods and introduce a maximum
of confusion in efforts to decipher the origins and environmental expansion of
the Amphibia. It is possible that many of these animals represent parts of
multiple transitions to land that ultimately failed, leaving no descendants.

Such confusion, and early suggestions that lungfish and amphibians were closely related, has led to many efforts to locate the source
or sources of amphibians among groups other than the rhipidistian
crossopterygians. Frustration and doubt about the efficacy and suffi-
ciency of the fossil record for determining relationships has led to
studies of living lungfish, amphibians (urodeles in particular), and
coelacanth crossopterygians (*Latimeria*) as the bases for character-by-
character analysis of these major groups. Out of this study of living
groups has come the proposal that lungfish and amphibians are the

closest sister groups among choanate vertebrates (but see essays in Schultze and Trueb 1991; Marshall and Schultze 1992). The patterns of evolution and development of terrestriality among early amphibians may be inferred from fossils of the late Carboniferous and the Permian. Both the temnospondyls and batrachosaurs developed branches that included distinctly terrestrial animals. Both groups have been found in the Scottish Visean site briefly described by Wood et al. (1985). These temnospondyls have well-developed tetrapod limbs, and they show some resemblances to dissorophoids, which include the most extensive array of terrestrial temnospondyls. The trend toward amphibian terrestriality appears to have developed very early. It was carried to an extreme in Permian dissorophoids such as *Fayella*. Whether or not such animals had retained the typical amphibian reproductive features is not determinable.

Batrachosaurs (anthracosaurs, s.l. of some authors) are prominent in Serpukhovian and later Carboniferous deposits and seem, as noted, to have been present in the late Devonian and early Carboniferous. Many were partially to completely aquatic, but in one or two clades terrestriality became well established. Notable in this respect are the seymouriamorph and the diadectomorph batrachosaurs. Among the former, aquatic larval stages existed during the Carboniferous (Spinar 1952), but whether this was the case for later Permian forms, such as *Seymouria* and *Diadectes*, is uncertain. That rapid changes in reproductive behavior occur in limited lineages is, of course, evident among various groups of the extant order Anura. In the case of *Seymouria*, faint grooves on the skull have been interpreted as remnants of lateral line canals. This may indicate an aquatic ontogeny, but the interpretation is at best uncertain. Recent studies by Berman et al. (1992) suggest that the Diadectomorpha are close to the amniotes.

The functionally terrestrial amphibians of the Permian appear to represent a radiation paralleling that of their reptilian contemporaries, but in general beginning somewhat later and never attaining the success of their counterparts. This imbalance is also seen today as the living anurans, while successful in terrestrial environments, in no way rival the successes of lizards and snakes or birds. What is more, these terrestrial amphibian lineages of the Permian all ultimately died out, leaving no apparent descendants. Reptiles, of course, are considered to have arisen among the amphibians, probably monophyletically. The most probable source is among the gephyrostegan batrachosaurs, but as considered next, any conjecture about origins and early terrestrial radiations produces problems.

6.5.2 Reptiles

Until very recently the first unequivocal record of reptiles had come from deposits of Moscovian age (late Carboniferous) of Joggins, Nova Scotia. At Joggins remains of both captorhinomorphs and pelycosaurs are present. The 1989 discovery in Scotland of a skeleton and skull of what appears to be a reptile put the age of the earliest known amniote back to the uppermost Visean (early Carboniferous). The age is estimated at about 340 mya, some 30 to 40 million years before that of the Moscovian (Smithson 1989). Possibly intermediate in age is *Romeriscus*, a purported reptile from the Bashkirian (Baird and Carroll 1967).

Validity of the Class Reptilia The matter of the origin of reptiles is closely tied to the question of whether or not the animals usually grouped under this term represent a valid clade or merely a grade— in other words, whether all three reptilian subclasses derive from a single common ancestor or do their characteristic adaptive features represent parallel or convergent evolution of distinct lineages with different anamniote ancestors. Some recent studies have recognized two distinct reptilian lineages: the Sauropsida, which includes some living reptiles and birds, and the Theropsida (or Synapsida), which includes the "mammal-like reptiles" and mammals. In most classifications, Aves and Mammalia have been retained as separate classes, as has Reptilia. Various other terminologies have been used, for the nomenclature is not stabilized.

The systematic position of the Captorhinomorpha, an order of the subclass Anapsida, is equivocal relative to the sauropsid and theropsid lineages. There is general agreement that the Captorhinomorpha comprises the most primitive known reptiles and, as such, contains the ancestry of the sauropsid and theropsid lineages (Carroll 1992). In some classifications it is included in the Sauropsida. One of its families, Captorhinidae, however, underwent an extensive radiation during the Permian and Triassic and perhaps may be considered as a third lineage. The base of all three could lie in the primitive captorhinomorph family Protorothyridae. Some light may be shed on this issue as the early Carboniferous "reptile" discovered in Scotland in 1989 becomes better known.

The question of monophyly of the Class Reptilia hinges upon whether or not there was a single ancestral species that meets all the criteria of "reptileness." Based on the commonly used criterion of reproductive modes, this comes down to the question of whether or not the amniote egg—the hallmark of higher vertebrate emancipation from aqueous reproduction—evolved more than once. The consensus has been and is that the complex structures of the amniote egg most probably could have arisen only once. Parsimony in phylogeny favors a single origin, but convergences arising within divergent lineages of organisms with equally complex structures are so extensive among both invertebrates and vertebrates that a multiple origin of such an egg is by no means out of the question. For simplicity the inclusive term reptile is used in the rest of this book without prejudice to one view or the other.

It is clear that the two major lines of development of reptiles (sauropsid and theropsid) have been separate from near the beginning of the reptilian fossil record. What is not at all clear is the time(s) of origin of the amniote egg. Were the earliest fossil animals we class as reptiles amniotes or not? Did an amniote egg develop within the gephyrostegans, the group of batrachosaurs that shows the greatest morphological resemblances to the very primitive reptiles? The simple truth is that neither the fossil record nor living animals provides clear answers to the question of the validity of the Class Reptilia. As is often the case, classification becomes a matter of convenience, expressing the phylogenetic views of its designer, but the actual phylogeny remains elusive.

The Environment of the Early Reptiles Problems of monophyly or polyphyly aside, the Carboniferous sites in which reptile remains have been found are informative about the nature and modes of life of early reptiles. The reptiles from the early Carboniferous of Scotland and the Moscovian of Nova Scotia are most instructive. The deposits in which the remains have been found appear to have been formed primarily under nonaquatic circumstances. The Scottish site has yielded an array of fossils, including myriapods, scorpions, a harvestman spider, temnospondylous and batrachosaurian amphibi-

ans, as well as the earliest known reptile. The reptiles at Joggins and
Florence, Nova Scotia, similarly are parts of faunal complexes that
are predominantly nonaquatic. The reptiles, and remains of other
organisms, occur in upright stumps of sigillarian trees. These animals
either accidentally fell into rotting tree stumps and were trapped
(Carroll et al. 1972) or the stumps were the actual places of habitation. The reptiles are associated with several genera of microsaurs,
small amphibians that were very similar in general habitus, and with
batrachosaurian and temnospondylous amphibians, which appear
to have been primarily terrestrial. Arachnids, insects, millipedes, a
probable arthropleurid, land snails, and *Spirorbis* annelid worm tubes
point, for the most part, to a land environment. The reptiles were
carnivorous and, based on their sizes, dentitions, and cooccurrences,
the arachnids and insects probably formed their principal food
source. The microsaurs, in part at least, probably had similar habits,
although some with rather blunt teeth may have had somewhat
different diets. As commonly seems to have been the case, once the
opportunity for exploitation of an environment becomes possible,
groups not necessarily genetically related will come to utilize the
emerging life zones. In this case, the speculation is plausible that
terrestrial environments lying close to shallow water and populated
with arachnids, insects, snails, and millipedes provided an opportunity for tetrapods to assume a fully terrestrial life.

However the initial thrust onto land may have taken place, it was
followed by a rapid adaptive radiation of the reptiles. This radiation
has some features that seem odd in view of the usual food webs with
which we are familiar in extant biotas. As far as the record shows,
very few strictly terrestrial, herbivorous reptiles or amphibians came
into being. Except for those small reptiles that fed primarily on invertebrates, the food sources of the various developing lines were dependent on aquatic organisms and, even though freed from the
reproductive necessity of water, they were still closely bound to it.
The relatively few large herbivorous forms that did evolve, which
were outnumbered many times over by the carnivorous and piscivorous amphibians and reptiles, also appear to have lived in or near
streams and lakes. This general trophic scheme is reflected in the
Carboniferous sites of Scotland and Nova Scotia and in the early
Permian sites from which a wide variety of tetrapods have come. The
first clear departure from this pattern, marked by the development

of an herbivore-carnivore reptilian trophic relationship, is found in faunas of the lower beds of the late Permian. That this new relationship evolved from the water-based systems common in the late Carboniferous and early Permian appears unlikely. It seems more probable that early radiations of reptiles that fed on insects and arachnids invaded drier and more upland areas during the late Carboniferous and persisted there, leaving evidence only in sporadic incursions into lowlands, until the beginning of the late Permian. At this time, the drier, more temperate conditions, which began to develop in the early Permian, became characteristic of the primary habitats for reptiles. This time marked the achievement of full emancipation from water by the tetrapods.

6.6 Summary

Few discoveries of early Carboniferous terrestrial metazoans have yet been made. A particularly important site in Scotland has produced associations of invertebrates and vertebrates that include myriapods, scorpions, the first harvestman, temnospondylous amphibians, and a probable representative of the earliest known reptile. Other sites in the midwestern and eastern United States plus Nova Scotia add to the sparse record. Tracheophyte plants continued their massive radiation, and interactions between plants and tetrapods as well as invertebrates appear to have been increasingly important phenomena. The Carboniferous as a whole marks the time during which initial occupations of the continents by metazoans gave way to terrestrial radiations of the principal groups. During this time and the subsequent Permian, continued continental movements and consequent mountain building led to replacement of warm, tropical climates by cooler, more xeric conditions, as the single continent of Pangea came into existence.

Among the terrestrial invertebrates arthropods dominated the continents. The radiations of the winged insects were spectacular, but each of the invertebrate groups that had become established by the beginning of the late Carboniferous tended to flourish either under partial or fully terrestrial circumstances.

At many localities in several present-day continents excellent preservation of fossils has allowed detailed taxonomic assignments of many of the fossils. These, in turn, suggest that in addition to the adaptive radiations, new invasions were taking place; polyphyletic adaptations to the land were thus probably the norm for all phyla. The first record of terrestrial gastropods has, for example, come from the Carboniferous, but it appears likely that other lineages of gastropods invaded the continents during the late Paleozoic and probably several times during the post-Paleozoic history of the earth.

The living semiterrestrial and terrestrial land leeches have no known fossil record. Circumstantial evidence indicates they may have arisen during the Mesozoic.

The fossil records of living groups of amphibious to terrestrial crustaceans (amphipods, isopods, brachyuran and anomuran decapods, crayfishes) are either unknown or fragmentary. The fragmentary records extend back only to the Cenozoic for some of the decapods, to the Mesozoic for the crayfishes.

Tetrapods assumed a prominent position among continental faunas during the late Carboniferous and the Permian. In contrast to the few known sites that carry amphibian remains in the early Carboniferous are the increasing numbers from rocks of the late Carboniferous and Permian. The discrete earliest occurrences of several major groups of amphibians pose serious problems about their origins and about the assumed monophyly of the amphibians as a whole. All of these animals, classified as amphibians, are by definition assigned to the Anamniota, but their reproductive and developmental pathways are largely inferential. The major radiations of the Paleozoic amphibians took place during the Permo-Carboniferous, with some lines continuing successfully into the first half of the Mesozoic era. High levels of terrestriality were attained by some of the lower Permian amphibians, approaching the extreme terrestriality of some of the Lissamphibia of today. However, there are no plausible connections between the Paleozoic groups of amphibians and living groups.

After the initial evidence of reptiles (presumed amniotes) in the early Carboniferous, the next records come from the middle part of the late Carboniferous. They are far from clear and pose many questions concerning the validity of the class Reptilia as commonly defined. Cladistic studies have emphasized many of the problems and have given rise to a revised nomenclature of the major "reptilian" groups.

The reptiles appear to have originated in terrestrial circumstances, but their predecessors among the presumed amphibian ancestors may have been either terrestrial or partially aquatic. The nature of the transitions from anamniotes to amniotes have been difficult to trace and to interpret. Initial radiations of reptiles were in terrestrial environments, with dependency on water only indirectly as a food source at the base of their trophic webs. Full emancipation from the water occurred during the early part of the late Permian. Throughout their history the reptiles underwent many radiations back into both freshwater and saltwater environments.

7

Functional Evidence from Living Molluscs

MALCOLM S. GORDON

The previous three chapters have summarized what is known and not known from the fossil record about the historical transitions from water to land for each of the major groups of metazoan animals that has successfully made that transition. We now turn our attention from the fossil record to living animals. In this and the following two chapters we summarize the present state of knowledge concerning the *functional* aspects of the transitions, as can be discerned from the major groups of living animals with fully terrestrial (as opposed to just cryptic) members: molluscs, annelids, arthropods, and vertebrates. As noted in chapter 1, we omit the insects from this part of the book because, as far as we know, there are no living forms that are primary occupiers of aquatic or transitional habitats. All known living aquatic or amphibious insects are secondary reinvaders of aqueous environments.

7.1 Scope of Study

The basic theoretical approach used here and in the next two chapters, and the main constraints and limits, are as described in chapter

1. There are four additional points to make relating specifically to the molluscs.

First, important starting points for this discussion are chapter 4, "Molluscs," in Little (1983) and the various sections on molluscs in Little (1990). Substantial amounts of additional relevant information are discussed in several chapters in the twelve-volume series *The Mollusca*, edited by W. D. Russell-Hunter and others, and published between 1983 and 1988. More recent primary publications are mentioned where relevant in the text that follows.

Second, there are no known fossil molluscs specifically identified as amphibious or otherwise directly related to water-land transitions. Multiple invasions of the land over long periods of time apparently were involved in the origins of living terrestrial molluscan groups. These facts combine to make it impossible to conclude whether living molluscs with physiologies and ecologies similar to what are construed to have been transitional strategies are, in fact, phylogenetically related to the forms that actually made early water-land transitions. Rather, they may have secondarily and convergently evolved life histories and life styles that mimic those transitions. Thus the two lines of indirect evidence—studies of phylogenetically related and unrelated living amphibious forms—described in chapter 1 are commingled in this chapter.

Third, most intertidal (eulittoral) molluscs, including all Polyplacophora (chitons) and Bivalvia and also the vast majority of gastropod orders, fall outside the scope of this chapter. These animals are aquatic; they are neither amphibious nor substantially terrestrial. Their tolerance for and survival of periods of emersion (even fairly long periods in many forms living in or above the upper intertidal zones) is based primarily on some form of air diving, usually breath-hold diving. They are largely inactive and withdrawn into closed or sealed shells. Those that air breathe are also mostly inactive. McMahon (1988, 1990) provides recent reviews of much of the literature on the physiology of eulittoral molluscs.

Fourth and finally, most pulmonate gastropods also fall outside the scope of this discussion, as most are completely terrestrial, apparently with very long evolutionary histories on the land (Solem 1985). We discuss terrestrial pulmonates only to the extent that specific features seem relevant to understanding the nature or process of the water-land transition. Living amphibious and aquatic pulmonates are generally considered secondary reinvaders of aquatic habitats, derived from completely terrestrial ancestors (McMahon 1983). Given the extreme polyphyly of pulmonates, however, this may well not be the case—at least not for all taxa. On this basis we do include information on a few amphibious pulmonates, both eulittoral and inland, as relevant to the water-land transitions of the group.

All the molluscan groups that meet the criteria we have established for this book belong to the class Gastropoda, subclasses Prosobranchia and Pulmonata. Little (1983) provides lists of the orders, superfamilies, and families that include amphibious and terrestrial species. There are about twelve such families among the prosobranchs and more than twenty such families among the pulmonates. The higher levels of classification within each subclass are currently uncertain and in a state of flux; thus the terminology used by Little requires significant revision. Graham (1985) and Solem (1985) give detailed discussions of the issues involved. All amphibious and terrestrial prosobranchs belong to superfamilies considered to be anatomically archaic (Graham 1985:169).

The striking evolutionary conservatism of individual gastropod taxa makes us hopeful that the functional properties of living amphibious and terrestrial forms may be fairly representative of the properties of their distant fossil antecedents. Whether or not this is actually the case, living amphibious gastropods provide us with envelopes of possibilities for each of the major features relevant to water-land transitions. Many variations on basic themes exist. Generalizations seem hazardous since present knowledge is based on a very small number of species.

Phylogenetically related aquatic, amphibious, and terrestrial gastropods are similar in many respects, including their basic body plans and such important functional features as foods and feeding patterns and mechanisms, bioenergetics, mechanical support and locomotor structures and mechanisms, and sensory mechanisms (Aldridge 1983; Kohn 1983; Trueman 1983; Graham 1985; Solem 1985; Carefoot 1987; Salvini-Plawen 1988; Trueman and Clarke 1988). In contrast to these conservative features, important differences relating to the water-land transition occur in: respiration and metabolism; thermal, water, and solute relations; waste metabolism and excretion; reproductive mechanisms; and some aspects of behavior and ecology. We will now consider in sequence each of the features in this latter group, to the extent the fragmentary and limited literature permits. Prosobranchs and pulmonates will be discussed in turn under each topic.

7.2 Respiration, Circulation, and Metabolism

There is a general morphological trend among amphibious and terrestrial prosobranchs, with numerous deviations, toward reduction and eventual elimination of gills and enlargement of mantle cavities in progressively more terrestrial forms. As gills are reduced, there is generally a compensatory increase in degrees of vascularization of parts or all of the walls of the mantle cavity. In some groups (e.g., some ampullariids) the mantle cavity is subdivided by a fold of tissue—one part holding the gill and used when the animal is in well-oxygenated water, the other with vascularized walls and used as a lung when the animal is either in poorly oxygenated water (by use of a siphon) or out of water. Smaller forms appear to depend primarily on diffusion for gas exchanges, while larger prosobranchs force ventilation by in-and-out movements of their heads (Aldridge 1983; Graham 1985).

Amphibious intertidal and inland pulmonates almost all use highly vascularized mantle cavities as lungs. These lungs are usually simple in structure. However, in at least one family of slugs (the stylommatophoran Athoracophoridae) the structure of the walls of the lung has become highly complex. A large number of fine, air-filled tubules, resembling insect tracheae in appearance, extend significant distances into the mantle tissues from the lung lumen.

In most basommatophorans and systellommatophorans the lungs apparently are ventilated primarily by gas diffusion. Stylommatophorans may also actively ventilate their lungs by moving the floor of the mantle cavity. Some high intertidal (eulittoral) pulmonates also have complex ciliated surfaces within their mantle cavities which function as secondary gills during periods under water (McMahon 1983; Pilkington et al. 1984; Morton 1988; McMahon 1988).

Few quantitative data are available concerning the respiratory role of the skin of the general body surface in either amphibious or terrestrial prosobranchs or pulmonates. Indeed, the literature on cutaneous respiration in molluscs is very limited (Graham 1988; Simkiss

1987). One group that must rely solely on cutaneous respiration is the terrestrial sluglike systellommatophorans of the family Veronicellidae. These forms have no mantle cavities. The amphibious intertidal systellommatophoran slug *Onchidium floridanum* (family Onchidiidae) respires cutaneously while submerged (Little 1983:52; Graham 1988:1040). Barnhart (1986a, 1992) indicates that cutaneous exchanges of carbon dioxide are important in the pulmonate land snail *Otala lactea*.

Amphibious prosobranchs and pulmonates generally appear able to respire equally well in either water or air. Most of the few forms studied have comparable oxygen consumption rates in both media (Carefoot 1987; McMahon 1988, 1990). Two pulmonates, *Melampus bidentatus* and *Cassidula aurisfelis*, and a prosobranch, *Cerithidea obtusa*—all three eulittoral fringe species (living actively above high tide lines)—appear, however, to do much better in air. Their metabolism becomes significantly anaerobic when they are in water (McMahon 1988).

The only substantial body of published data for amphibious gastropods that describes the relationships between aerobic metabolic rates and a range of ecologically and physiologically important abiotic and biotic variables relates to a saltmarsh pulmonate, *Melampus bidentatus*, that is a eulittoral fringe dweller of the temperate zone (McMahon and Russell-Hunter 1981). In addition to being more aerobic out of water, these snails show strong circadian rhythms in oxygen consumption and relatively low thermal sensitivities (Q-10s) for oxygen uptake, and they demonstrate a relatively unusual pattern of metabolic acclimation to temperature ("reverse acclimation").

There appear to be no consistent, substantive differences in the structure or the functioning of the circulatory systems of amphibious or terrestrial gastropods and those of phylogenetically related aquatic gastropods (Jones 1983). Studies of the circulatory systems of extant gastropods, therefore, offer very limited help for understanding the water-land transition. There is some possibility that terrestrial pulmonates have more nearly closed circulatory systems that operate at higher blood pressures, but the data base is inadequate to support such a generalization (Jones 1983). Based on the more extensive vascularization of the walls of mantle cavities in more terrestrial prosobranchs, perhaps the patterns of blood flow in these forms may be somewhat different from those in more aquatic forms.

An essential component of the suite of functional capacities that together permit the survival of both amphibious and terrestrial gastropods is their tolerance of periods of environmentally adverse conditions—whether too hot, too cold, or too dry. Survival is generally ensured by the animals becoming dormant for the duration of the difficult time. Diurnal dormancy is widespread, especially in terrestrial groups. Extended, often seasonal, dormancy also occurs widely. In some desert areas extended dormancy can be continuous for as long as five years (Burton 1983).

Behavioral and physiological mechanisms involved in dormancy vary considerably among different gastropod groups. There are no apparent uniform phylogenetic or environmental correlations with respect to which group does what. Some forms burrow into mud or soft soil and then seal up their shells. Others secrete mucous barriers around shell openings (epiphragms), which bond them firmly to their substrates (e.g., under rocks or downed logs and also sometimes in exposed situations in the eulittoral fringe), and then withdraw some distance back into their shells. All become inactive during this time. The few forms that have been studied in some detail while dormant for extended periods (especially the pulmonate helicids *Helix* and *Otala*) show rates of oxygen consumption averaging much lower than rates normal while active at similar temperatures. These low rates are partly due to lowered lung ventilation (Barnhart 1986a).

Dormant snails regulate internal acid-base balance by varying and controlling both oxygen uptake from and carbon dioxide release to the atmosphere. Their heart rates are greatly reduced. Lung hypoventilation in helicids results in carbon dioxide retention and acidosis (Aldridge 1983; Barnhart and McMahon 1987, 1988; Barnhart 1986b, 1989, 1992; Rees and Hand 1990). Brooks and Storey (1990) describe important features of the biochemical adjustments underlying these physiological changes.

Periods of dormancy are ended when environmental conditions return within limits tolerable to the particular species. For diurnal dormancy in terrestrial forms arousal usually occurs at or near nightfall, as air temperatures drop and humidities rise. For extended periods of dormancy arousal can occur gradually, as with spring warming, or very suddenly, as with the first heavy rain of a rainy season or the high spring tides. The mechanisms involved in molluscan arousals from dormancy are essentially unknown, though many stimuli cause dormant snails to arouse, oftentimes surprisingly rapidly (Barnhart and McMahon 1987:135; Barnhart 1989).

Exposure to subfreezing temperatures is a particularly difficult environmental challenge for many amphibious and terrestrial molluscs. It seems likely that the geographical limits of the poleward distributions, also the upper altitudinal limits, of amphibious and terrestrially hibernating gastropods are substantially related to their abilities to resist or tolerate some degree of freezing of their body fluids and tissues during winter periods. Metabolic rates under these circumstances are greatly reduced. Daily cycling of metabolic rates occurs at least in those temperate zone and subarctic eulittoral prosobranchs and pulmonates (some of which are amphibious) that freeze and thaw daily (often twice daily) during winters as tidal heights change (Hilbish 1981; Aarset 1982; Murphy 1983; Storey and Storey 1988). Freezing resistance by supercooling seems to be the primary response of inland amphibious and terrestrial gastropods (Riddle 1983).

7.3 Thermal, Water, and Solute Relations

Temperature as a major environmental variable has, by and large, not received much attention in the ecological and physiological literature on amphibious gastropods. Some work has been done on the bases for tolerance of subfreezing temperatures. Excepting only some recent work on eulittoral prosobranchs, there is a notable lack of information on such standard thermal concerns (in other animal groups) as thermal tolerance limits, temperature preferenda, and behavioral thermal regulation. There may thus be much of interest to be learned.

Upper thermal tolerance limits and aspects of thermoregulation resulting from evaporative water losses have recently been studied in a number of amphibious eulittoral and eulittoral fringe prosobranchs in Hong Kong and western Australia (McMahon and Britton 1985; Cleland and McMahon 1990; McMahon 1990; McMahon and Cleland 1990). Snails representing a number of different families were studied in both rocky shore and mangrove environments. Contrary to results from studies of nonamphibious intertidal molluscs, there was no correlation between upper limits of thermal tolerance (heat coma temperatures) of amphibious forms and the vertical zonation positions of the species in the eulittoral. This lack of correlation apparently was due to a combination of physiological and behavioral capacities of the snails. Many species use rapid rates of evaporative water loss from exposed body surfaces to maintain body temperatures well below ambient levels that often exceed their tolerance limits. These species are tolerant of both rapid rates of water loss and of substantial total amounts of water loss. However, when desiccation reaches an upper limit they become inactive and withdraw into their shells, thus greatly slowing further desiccation. Behaviorally, many species select sheltered, moist microhabitats (e.g., rock crevices or areas covered with algal mats) and thus avoid the direct effects of high ambient temperatures. Those few eulittoral fringe species that remain active when emersed may have somewhat higher upper thermal limits than the eulittoral species, but not as much higher as might be expected. Here again behavioral selection of cooler, moister microhabitats plays an important role.

The most spectacular performance in inclement conditions is perhaps that of the mangrove snail *Cerithidea ornata*. This species lives in Hong Kong on the trunks of mangrove trees well above the ranges of all other gastropods and well above highest high tide levels. It appears totally aerial in its habits. Its upper thermal tolerance is no higher than values for many species living much lower in the eulittoral. Like many eulittoral species it is tolerant of substantial degrees of dehydration, and it uses evaporative cooling during activity periods as long as five days without contact with water. After five days it becomes inactive and greatly reduces its rate of evaporative water loss. Many individuals in the dormant state almost completely cease body weight loss, suggesting highly efficient mechanisms for water conservation.

The water relations of other groups of amphibious prosobranchs are mostly unstudied. Partial exceptions to this statement are a few species of ampullariids, cyclophorids, and pomatiasids (Aldridge 1983; Little 1983, 1990) for which some information is available on hemolymph concentrations and compositions and on some aspects of waste production and excretion.

The water relations of amphibious and terrestrial *pulmonates*, in contrast, have been studied much more extensively. Several features of these relations appear to be so general in their occurrence that chances are excellent that they represent major features of the water-land transitions for all gastropods (Burton 1983; McMahon 1983; Riddle 1983; Prior 1985; Little 1990). Perhaps the most basic feature is that, as far as is known, all amphibious and terrestrial pulmonates have integuments that are as permeable to evaporation of water as free water surfaces of the same area. They are, in this respect, the same as most vertebrate amphibians. The major consequence of this condition, of course, is that, when out of water, all pulmonates are highly susceptible to desiccation as a result of cutaneous water loss. Additional water losses are sustained both as a result of respiratory evaporative water loss and as part of the production of the dilute mucus used in locomotion. The simultaneous operation of these three pathways for water loss can lead, even under conditions of moderate environmental humidities, to rates of water loss from medium-size snails as high as 15–20 percent of initial body weight per hour. Very small snails may lose water extremely rapidly; thus they are likely to be cryptic.

Associated with this basic lack of control over water loss is a high tolerance for large changes in levels of body hydration. Some pulmonates have been found to tolerate total water losses as high as 80 percent of body weight (McMahon 1983; Prior 1985).

Comparably large changes occur in the osmotic and major solute concentrations of both the circulating body fluids (hemolymph) and the intracellular fluids. Proportionate losses of water vary among different organs and tissues, the animals apparently exerting substantial differential regulatory control over water content of some organ systems as compared to others. But the hemolymph sustains the greatest proportional water content changes.

Even though only a few different species of terrestrial pulmonates have been studied with respect to intracellular osmotic and solute regulation—and even fewer with respect to such functional consequences of these concentration changes as changes in important aspects of intermediary metabolism, neuronal excitability, and synaptic transmission—the complexity of the phenomena and processes occurring is substantial. Multiple solutes are involved, both inorganic ions and various free amino acids. Most, if not all, of these phenomena and processes can be legitimately considered to be molecular, biochemical, cellular, tissue, and organ level components of the water-land transition in the molluscs. An effort to summarize these mechanisms in any detail here would, however, divert attention from our major focus at the organismic level. We therefore refer interested readers to the reviews of Burton (1983), Prior (1985) and Somero (1987) for further information.

Terrestrial gastropods respond organismically to the physiological challenges just described in two major ways. First, they vary their behaviors so as to minimize, or at least to reduce to tolerable levels, their water losses. Most forms do this by restricting their activities to moist or humid habitats; in more xeric environments, they become nocturnal. The final behavioral response to desiccation nearing tolerable limits is to become dormant, either on a daily or a more extended basis (McMahon 1983; Barnhart 1989). Second, they have developed capacities to rehydrate themselves rapidly whenever presented with an opportunity. Their high cutaneous water permeability, combined with the greatly lowered water potentials of their body fluids when they are dehydrated, enable them to absorb water from even moist surfaces. Many forms have evolved complex behavioral patterns and responses that facilitate rehydration. These are described by Prior (1985).

The osmotic and solute concentrations of the hemolymph in fully hydrated states form the baselines to which amphibious and terrestrial gastropods ultimately try to return. Here again, however, diversity is the most striking feature. Assuming that the small numbers of species that have been studied are to some degree truly representative of physiologies throughout each of the major groups, all that can be said is that baseline osmotic concentrations usually vary from low to moderate (60–270 mOsm/l), with sodium and chloride as the two major monovalent inorganic ionic constituents. Amphibious eulittoral fringe species living in estuarine mud flats and salt marshes show a variance in baseline values of much wider ranges, depending on their conditions of salinity adaptation; they are thus osmoconformers (Little et al. 1984).

There do not appear to be any special features of hemolymph concentrations or compositions that are related uniquely to waterland transitions. In fact, the standard interpretation of the diversity in living gastropods is that the differences reflect the evolutionarily ancestral environments (marine or fresh water) of the various groups, rather than being specific adaptations to current environments or habitats (Burton 1983; Andrews 1987; Little 1990). Our thought on this is: "perhaps."

Dormancy has been mentioned several times as the last-ditch strategy used by gastropods faced with continuing, severe dehydration. The diversity of refugia used and of physical mechanisms involved in entry into dormant states was described earlier. The primary water-related consequence of entry into dormancy appears to be a substantial reduction in rates of overall water losses—often to only a small percentage of the rates occurring in active animals. The bases for these large reductions also vary between groups. The pattern of variations in water conservation mechanisms shows no apparent correlations with important aspects of the water-land transition.

Since dormant snails cannot fully arrest either metabolism or water loss during extended periods of dormancy, this response to adverse conditions has substantial consequences for body water contents, internal distributions of water, and fluid concentrations and solute compositions. Once again only a few forms have been studied in these respects, but both prosobranchs and pulmonates were among them (Burton 1983; McMahon 1983; Riddle 1983; Barnhart 1986b). A common pattern in snails regularly undergoing seasonal dormancy is a gradual increase in hemolymph osmotic and major ionic concentrations during dormant periods. These concentrations may reach levels several times higher than baseline concentrations in normally hydrated, active animals. Concentrations also tend to be highest at the end of dormant periods, and return to baseline levels occurs relatively rapidly once activity is resumed. The overall magnitudes of these changes are comparable to the changes occurring over much shorter time periods in active gastropods encountering difficult environmental conditions.

A final interesting aspect of the solute-related ecophysiological adaptation of terrestrial gastropods (though it is not unique to them) relates to their needs for substantial amounts of calcium and magnesium salts, especially carbonates and phosphates. This is the case for both shelled and unshelled forms; most snails, with or without shells, accumulate significant amounts of divalent ion-based mineral deposits in various tissues (Burton 1983; Carefoot 1987). Their requirements for these minerals appear to be one important factor determining geographic distributions. They do best in regions with calcium- and magnesium-rich surface waters, or with limestone or similar rock formations. This requirement is clearly not a new evolutionary development that was part of the water-land transition.

Rather, it is an apparently important restriction on the types of environments in which the water-land transitions could take place.

7.4 Waste Metabolism and Excretion

The structure and function of gastropod excretory systems remain among the major research interests of molluscan physiologists, functional morphologists, and systematists. As a result, the data base on this system relating to amphibious and terrestrial gastropods is both more detailed and somewhat more phylogenetically diverse than for many of the other systems we have already considered (Martin 1983; Andrews 1987; Little 1990).

Gastropod kidneys, both prosobranch and pulmonate, differ dramatically from vertebrate kidneys in their structures, but they apparently function very similarly. With a few possible exceptions (e.g., a tubular gland in the foot of pomatiasids, which may be important in hemolymph osmoregulation as a pathway for uptake of inorganic ions from soil water), kidneys are the primary organs used by amphibious and terrestrial gastropods for excretion of metabolic wastes and for regulation of hemolymph ionic compositions. Thus kidneys must have played important roles in water-land transitions in these groups. Just what these roles were, however, is unclear.

Given the susceptibility of gastropods out of water to rapid and severe dehydration, one might expect that when snails are on land (1) rates of urine production should be substantially lower than those found in the same snails in water, and (2) urine osmotic concentrations should be as high as possible—at least isosmotic with hemolymph, if not higher. What do the data show?

Rates of urine production by either prosobranchs or pulmonates out of water are highly variable and may be quite low, but often are comparable with those of snails in water. Many snails appear to be easily stressed as a result of handling, such as occurs in cannulations of urinary pores; variability in levels of body hydration in snails used in experiments may also be a factor. Urine osmotic concentrations in a few marine eulittoral littorinids and some inland amphibious pomatiasids are close to, if not exactly, isosmotic with hemolymph concentrations. Urine osmotic concentrations in all other amphibious and terrestrial snails that have been studied are usually strongly hypoosmotic to hemolymph.

The conventional wisdom with respect to these results is that requirements for solute conservation are more severe than for water conservation. Thus proportionately larger amounts of salts and other important solutes (e.g., glucose) than water are reabsorbed from hemolymph filtrates. This conclusion seems to us not well founded. The relevant data base is inadequate. More rigorous demonstrations are necessary to confirm that proportional rates of loss of important solutes are on average higher than are proportional rates of water loss. Experimental evidence is also needed that overall rates of solute intake (from food, ambient water, or by cutaneous uptake) average less than rates of solute loss.

A great deal of attention has been paid both to the chemical forms of waste nitrogen produced by and usually excreted by amphibious and terrestrial gastropods and to the anatomical sites and physiological processes used in this excretion. One of the major concerns in such work has been the problems associated with extended periods of dormancy. It seems clear, however, that essentially all gastropods, both aquatic and others, possess the biochemical machinery needed to produce the full spectrum of waste nitrogen end products. Amphibious and terrestrial forms may differ quantitatively from the others, but not qualitatively (McMahon 1983; Carefoot 1987).

Once again variability is a major feature of the findings. The numbers of species studied in each of the major groups, both of prosobranchs and pulmonates, is very small, so there are the usual questions of both the accuracy of representation of the group's overall properties and the reliability of possible correlations that may be perceived with other processes or phenomena—like the water-land transition. Our perception is that the within-group variabilities are so large that nothing reliable can be said about between-group differences.

Proceeding on this basis, it is clear that amphibious and terrestrial gastropods have available to them several options for disposing of waste nitrogen. A common strategy is to eliminate variable, but often substantial, amounts of nitrogen as ammonia gas. This is apparently most often accomplished by simple diffusion across the walls of the mantle cavities. A second common strategy is to form concretions from various purines, especially guanine and xanthine in addition to uric acid and its divalent ion salts, along with calcium salts and such other components as mucopolysaccharides. In some taxa, possibly the short-lived species being most frequent, these concretions are simply accumulated in the tissues. If the tissue is the kidney, the kidney may be considered a kidney of accumulation. Most other snails, however, excrete these concretions. At least among the species of amphibious and terrestrial gastropods that have been best studied, urea seems not to be an excretory form of choice under most conditions. Some amphibious basommatophorans are exceptions to this statement (McMahon 1983). Relatively low concentrations of urea have been measured in the urine of some snails from time to time, especially during active feeding when ample water is available. Higher urea concentrations may accumulate in some forms during dormancy (Barnhart 1989).

Andrews (1987) provides a thorough and detailed review of many ultrastructural and microstructural features of the kidneys and associated excretory structures in amphibious and terrestrial gastropods.

7.5 Reproduction

The water-land transitions of both amphibious and terrestrial prosobranchs and pulmonates seem to have required little or no evolutionary innovation with respect to general strategies for reproduction. Since unequivocal ties to the fossil record are lacking, there is no way we can determine whether or not the reproductive patterns and mechanisms used in extant groups of snails derived from preexisting genetic instructions in their aquatic ancestors. All that can be said is that virtually the full spectrum of reproductive patterns and mechanisms known from extant marine and freshwater gastropods also occurs in extant amphibious and terrestrial groups (Little 1983, 1990; Fretter 1984; Tompa 1984; Audesirk and Audesirk 1985).

Many amphibious snails, both eulittoral and inland, return to aquatic habitats to reproduce. These forms release into water or deposit on submerged surfaces eggs—many of which were fertilized internally—that hatch into planktonic larvae. Other amphibious species are either ovoviviparous or truly viviparous, giving birth in aquatic surroundings to fully developed small snails.

Fully terrestrial snails universally (or nearly so) have internal fertilization, direct development without free-living larval stages, and shelled eggs. Prosobranchs generally have the sexes separate; pulmonates are hermaphroditic. Mating systems are extremely complex and varied. Both functional protandry and protogyny occur, also simultaneous hermaphroditism. Fertilization may involve matings (either one-way or reciprocal) or self-fertilization or parthenogenesis. Some forms lay eggs; others retain fertilized eggs for some period before laying them; still others are ovoviviparous or even viviparous. Eggs produced by egglaying forms vary widely in numbers, size, and structure. Shells vary from relatively leathery in texture to fully calcified and looking like small bird eggs. Some species lay eggs in groups, each group containing one fertile egg with the others infertile and used by the hatchling as food.

Given the extremely long evolutionary periods that many of the terrestrial groups have been on land, combined with the very limited geographic mobility of most forms, this diversity in reproductive strategies is not surprising. However, it is impossible to determine how much of this diversity, if any, had direct associations with water-land transitions.

7.6 Behavior and Ecology

Excepting only a few descriptions of field behaviors summarized by Little (1990), the behavioral ecology of amphibious snails—especially in environmentally relevant contexts that would be most helpful for understanding the water-land transitions—is almost unknown. The literature is almost devoid of information relating to these animals, either prosobranch or pulmonate, with respect to even such basic issues as preferences for land or water, also ecologically important choices of microhabitats in both environments.

This gap is all the more striking since the study of the neurobiology of gastropod behavior in many other contexts (from sensory responses to central nervous system information processing and learning) is a very active, rapidly expanding, and sophisticated field of research. Much of this neurobiological research relates to internal mechanisms, but a substantial fraction is ecologically related. A few amphibious prosobranchs and pulmonates have been used in some of this work, but the vast majority of the effort involves aquatic and terrestrial pulmonates, plus various opisthobranchs (see all four chapters in Willows 1985).

The water-land transitions of gastropods necessarily involved evolutionary innovations in the areas of sensory physiology and related behaviors. Among these must have been, in at least some groups, quantitative if not qualitative changes with respect to responses to ambient and internal temperatures, water losses, metabolic rates, and responses to light and odors. For each of these parameters both ranges of variables and rates of changes are certainly different in aquatic as compared with terrestrial habitats. The natures of the stimuli also change, especially with respect to light and odors. Appropriate behavioral responses to these differences should also vary between the two sets of habitats.

Substantially more is known about the physiological ecology of both inland amphibious and terrestrial prosobranchs and pulmonates (Aldridge 1983; McMahon 1983; Riddle 1983; Little 1990). We have already summarized the main features of this work relevant to water-land transitions.

Large differences exist between many aquatic and terrestrial gastropods with respect to important features of population ecology, population genetics, and life history patterns (Cain 1983; Calow 1983). There is much more opportunity for gene flow within and between fairly widely distributed and separated populations of marine forms having planktonic larvae than there is between populations of terrestrial snails having direct development. Terrestrial snails (pulmonates are best studied) frequently have panmictic populations restricted in size to only a few hundred individuals occupying a territory of only a few hundred square meters. The literature relating to the population ecological and genetic consequences of each of these different life history patterns is substantial, but none of this work appears directly relevant to water-land transitions.

7.7 Summary

The main points relating to water-land transitions among the gastropods are:

1. There are many kinds of living amphibious and terrestrial prosobranchs and pulmonates. Very few of these have been studied at all, and even fewer have been studied with respect to the water-land transition.

2. As is the case for most molluscan groups, amphibious and terrestrial snails have been evolutionarily conservative. They are very

similar to, if not indistinguishable from, their closest aquatic relatives with respect to such important features as basic body plans, foods and feeding patterns and mechanisms, bioenergetics, mechanical support and locomotor structures and mechanisms, and sensory mechanisms. Because these features have not changed in the course of transitions to the land, they may be regarded as preadaptations for terrestriality.

3. At the same time a variety of major functional morphological, biochemical, physiological, behavioral, ecological, population dynamical, population genetic, and life historical changes have occurred in most, if not all of the taxa that successfully made the water-land transition. The diversity of adaptive patterns and combinations of patterns is large.

4. A great deal of interesting and important work remains to be done relating to all aspects of the transitions. In particular, the dearth of information relating to embryonic development and early life history stages is a substantial impediment to understanding how the transition to land might have occurred.

5. It is not yet possible to develop any credible scenarios concerning how and where any of the amphibious or terrestrial groups of living gastropods first made their transitions to land. The probability is high that there were many transitions in each major group, in many places, in varied habitats and environments, over long periods of time. Many of the living amphibious forms appear to be secondary reinvaders of aquatic habitats.

8

Functional Evidence from Living Annelids and Crustaceans

MALCOLM S. GORDON

In this second of three chapters exploring what can be learned of the water-land transition from functional evidence available in living metazoans, we focus on two phyla: Annelida and Crustacea. Among extant annelids, the land leeches are the only non-cryptic terrestrial forms. Among the crustaceans, there are five groups containing amphibious to terrestrial forms: amphipods, isopods, crayfishes, and brachyuran and anomuran crabs.

8.1 Scope of Study

The principal starting points for this discussion are the sections on leeches and crustaceans in the 1983 and 1990 books by Colin Little. Additional relevant information is presented by Sawyer (1986) and Burggren and McMahon (1988).

We consider the various species of land leeches to be the only annelids that meet our criteria for designation as amphibious to fully terrestrial—although many do require humid environments. The other annelids are all cryptic organisms, meaning they are still primarily aquatic. We note in passing that Sawyer (1986) makes a strong argument that the members of the class Hirudinea, the leeches, are not really annelids, though their similarity to oligochaetes is not questioned. Sawyer considers the Hirudinea to be substantially "arthropodized" and thus closer to the phylum Uniramia than to any other major systematic category.

The groups of leeches that have evolved fully terrestrial life histories are all included in the subclass Euhirudinea. There are two orders in that subclass: Rhynchobdellida and Arhynchobdellida. Virtually all terrestrial leeches are arhynchobdellids and are found in three families: Cylicobdellidae and Americobdellidae are predaceous, whereas the Haemadipsidae are haematophagous—feeding on blood. Many species in other families of this same order are more or less amphibious (see Sawyer 1986).

As to the phylum Crustacea, the fully terrestrial representatives are currently placed in the class Malacostraca. The superorder Peracarida includes the isopods (order Isopoda) and the amphipods (order Amphipoda). The superorder Eucarida includes the decapods (order Decapoda), which are further subdivided into the two suborders of crabs (Brachyura and Anomura) and the suborder of crayfishes (Astacidea). All these groups contain a few to many amphibious to fully terrestrial members. (Schram [1986] presents a somewhat different classification of the Malacostraca.)

What about the other two phyla of arthropods? Since almost all known living aquatic or amphibious chelicerates (excepting xiphosurans) and all such uniramians are, without doubt, relatively recent secondary returnees to those lifestyles, we omit these two phyla from further discussion (Messner 1988).

8.2 Land Leeches

Land leeches are virtually unknown to most residents of North America or Europe, including many trained biologists in these regions. They are, however, widely distributed in much of the rest of the world and often occur in substantial numbers. Adults of many species reach large sizes, commonly 5–10 cm length even when at rest and unfed.

Each of the three families of land leeches is widely distributed. The Cylicobdellidae occur in South America, east Asia, and many of the Pacific islands, including the Hawaiian islands. The Americobdellidae are South American. But it is the blood-sucking family, the Haemadipsidae, that is by far the most widely distributed, occurring in southern South America and Africa, south and southeast Asia, Australasia, some southern European countries, and various islands in the western Indian Ocean, including Madagascar. The populations on isolated oceanic islands are considered to have derived from leeches carried by birds (Sawyer 1986).

8.2.1 Functional Morphology and Locomotion

There are a few significant morphological differences between terrestrial and nonterrestrial (aquatic or amphibious) members of the three families. Thorough comparative studies have not yet been made, but at least a few species of haemadipsid land leeches have modified structures for some or all of their nephridia. The urinary bladders are enlarged, and the openings of the terminal nephridiopores are positioned differently from those of aquatic forms, whose openings are typically located ventrally. In the land leeches the nephridiopores are at least lateral and often dorsal to the lateral margins of the animals. These excretory modifications of the haemadipsid land leeches are considered adaptations to keep the general body surfaces of the animals moist. In addition, in some species either the most anterior or both the most anterior and the most posterior nephridiopores open near the margins of the oral and/or caudal suckers. These arrangements probably help form aqueous seals around the margins of the suckers when the animals wish to attach themselves to something (Sawyer 1986:632).

The suckers of at least some species of haemadipsid land leeches differ structurally from those of aquatic leeches in several respects. The caudal suckers especially are relatively large and heavy. They are usually elliptical in outline, rather than circular, and terminate anteriorly in pointed triangular lobes called prehensile papillae. These papillae operate somewhat in the manner of opposable digits, permitting the animals to grasp objects (such as small twigs or grass stems) in ways that are not possible for most aquatic leeches. Many muscular ridges are also present on the inner surfaces of the suckers that probably assist in attachments to objects with dry surfaces.

These modified suckers are strong in themselves, and their strength is supplemented with a uniquely structured and strengthened dorsoventral system of muscles. This system is made of groups of straplike muscles that originate postero-ventrally and insert antero-dorsally. The details of the systems vary somewhat interspecifically, but they serve to permit the animals to move strongly, agilely, and quickly over rough substrates. The primary mode of locomotion is inchworm-type crawling. This locomotory mode is also used when land leeches are in water (which they normally avoid, if possible). It is not certain, however, whether they are absolutely incapable of swimming, which is the primary locomotory mode of most aquatic leeches, but land leeches have not been observed to swim (Sawyer 1986:383).

Traveling land leeches that pick up bits of debris on either sucker as they move along will remove those objects by rubbing the sucker against something else, most often their own ventral surfaces. Aquatic leeches normally would not encounter this problem, and appear to lack this behavior pattern if the problem does occur (Sawyer 1986:632–33).

8.2.2 Ecophysiology, Behavior, and Ecology

The primary food sources for most land leeches are mammals and birds (they occasionally feed on the flesh or blood of amphibians and reptiles as well). There are two major sets of physiological problems associated with such feeding habits. First, leeches must be able to position themselves in locations where their food sources are likely to pass by, and they must be able to accurately and rapidly locate those sources when they are present. Thus an array of sensory physiological adaptations must be present that permit them to accomplish this essential task. Second, they must be able to successfully deal with a wide range of conditions relating to water balance. Even though land leeches are restricted primarily to humid environments, their integuments are quite permeable to evaporative water losses. Thus, while waiting in suitable locations for a passing animal they are likely to suffer variable, often considerable, amounts of desiccation. In addition, once they have successfully attached themselves to an animal, the haematophagous forms in particular must deal with sudden large fluid loads.

Ecophysiological, behavioral, and ecological understanding of how land leeches deal with these challenges is in a primitive state. Very few species have been studied at all—most of those in fragmented, partial, and relatively simple ways (Sawyer 1986).

Water availability is the single most important factor affecting local and seasonal abundances and distributions of land leeches. Where the environment is almost continually moist, these animals are active throughout the year. Areas having distinct wet and dry seasons show marked seasonal variations in abundance of active animals. They are often extremely numerous in agricultural areas with many domesticated farm animals, such as in northern India. During the dry seasons in these areas active leeches are rare, but large numbers may be found burrowed underground near watercourses, attached to the undersides of buried stones. These animals are largely inert and unresponsive to stimulation. This hibernation-like state has been given the name *anhydrobiosis*. When the rainy season begins these leeches quickly revive and return to activity on the surface.

Diurnal activity cycles in land leeches appear to result from interactions between at least three different factors: (1) whether the species is predaceous or haematophagous; (2) light levels—predaceous leeches are generally more photonegative than are haematophagous leeches; and (3) ambient humidity variations. Species that live primarily on or in moist litter on the forest floor (ground leeches) often show two daily activity periods, morning and early evening. These species are more likely to be haematophagous. Other, often closely related species, climb into and wait in low-lying bushes or lower limbs of trees (bush leeches). Bush leeches are more likely to be predaceous and are most active in early morning hours before sunrise.

Absent excessive exposures to bright light, high temperatures, or rapid air movements even aquatic leeches can tolerate long periods out of water and substantial desiccation (up to 80 percent of body water content in a few forms that have been studied). Land leeches are thought able to do at least as well, but no data are available to make precise comparisons. The retention by both aquatic and land leeches of ammonia as the major end product of nitrogen metabolism (about 80 percent of total waste nitrogen in one species, the rest being urea) is thought to be a major factor restricting land forms to damp environments. (Only one species of land leech has been studied in this connection. It did not excrete any uric acid.)

Reliable data are not available with respect to essentially all other aspects of land leech ecophysiology. Their successful occupation of a wide variety of habitats (e.g., some species have altitude ranges from near sea level to in excess of 4,000 meters) indicates there is much to learn relating to effects of temperature, oxygen partial pressure, and ultraviolet radiation.

Qualitative observations of behavioral responses of a few species of land leeches indicate that their sensory physiology is complex. They share with many aquatic leeches apparent heightened states of alertness in response to moving shadows. Aquatic leeches also respond to water movements in their vicinity; land leeches respond to air movements. There is some indication that land leeches may be attracted by airborne odors that might indicate the presence of prey organisms, and they also have photoreceptors around the edges of their oral suckers that may possess some sensitivity to infrared radiation. Ground leeches almost certainly can detect and respond appropriately to sound or vibrations transmitted through air or the substrate that might indicate the presence of prey.

Reproduction in land leeches appears to occur by mechanisms similar to those used by aquatic and amphibious forms. Eggs are often laid beneath the surface of moist soil, enclosed in structures called cocoons.

In summary we can say that, despite general omission from most previous discussions of terrestrial invasions by animals, land leeches are indeed successful invaders of the land by all relevant criteria. Science is ignorant of many significant aspects of their adaptations to terrestrial life.

8.3 Amphibious and Terrestrial Crustaceans

Little (1983) points out that only about 5 percent of known crustacean species are amphibious (semiterrestrial) or terrestrial; this amounts to somewhat fewer than 1,500 species altogether. Most of these, about 1,000 species, are fully terrestrial isopods. Thus the actual taxa of crustaceans that may provide some relevant indirect evidence con-

TABLE 8.1
Grades of Terrestrial Adaptation in Crustaceans

Grades	Characteristics
T-1	Resident in fresh or salt water; active primarily in water, but briefly and intermittently active in air.
T-2	Intertidal species which usually conceal themselves and are inactive while covered by water. Only active in air during low water periods, diurnally, nocturnally, or both.
T-3	Resident littorally, inland, or out of fresh water; active in air and usually nocturnal. Most often fossorial or cryptic. Require regular access to water for immersion, either in water bodies or at burrow bases. Dependent on water for pelagic larval stages.
T-4	As in T-3, but not requiring regular immersion. Can obtain metabolic and preformed water from food, can drink dew or other surface water, and can take up water from damp substrates. Dependent on water for pelagic larval stages.
T-5	As in T-4, but with larval development either shortened or direct.

SOURCE: After Powers and Bliss 1983 and Hartnoll 1988

cerning functional adaptations for water-land transitions, within the policy limits we have set for this book, represent a tiny fraction of crustacean diversity. An even smaller subset of forms has been studied in any detail in these contexts. We therefore prefer not to speculate concerning which, if any, of the forms that have been carefully investigated may be the more relevant to how the actual processes of transition might have taken place. Present knowledge simply provides a set of envelopes of possibility.

Little (1990) presents an array of intriguing and plausible scenarios concerning the possible environments in which the various terrestrial crustacean groups may have arisen and the more specific habitats in which these groups made their transitions to land. We view these scenarios as useful springboards for further studies.

A striking general feature of virtually all living amphibious and terrestrial crustaceans is how little they differ structurally from their nearest apparent aquatic relatives. The basic *bauplans* of the various groups have been successfully adapted to function in subaerial environments with only the most minimal macroscopic modifications. This morphological conservatism parallels the situation in the amphibious and terrestrial molluscs and has comparable consequences for this discussion. It is undoubtedly a major reason why paleontologists have so far been unable to definitively identify crustacean fossils relating to water-land transitions. It also clearly requires that essentially all important changes that must have occurred during these transitions were functional in nature, rather than structural. Therefore the rest of this consideration of crustacean transitions emphasizes biochemistry, functional morphology, ecophysiology, behavior, and ecology.

Large amounts of supplementary and complementary information on almost all aspects of crustacean biology may be found in the ten volumes of *The Biology of Crustacea*, edited by D. E. Bliss and collaborators in 1982–85. Powers and Bliss (1983) provide a general summary of terrestrial adaptations. The recent symposium edited by Mantel (1992) provides the newest information. Table 8.1 summarizes a classification of grades of terrestrial adaptation developed primarily for land crabs, but also applicable to other amphibious and terrestrial crustaceans (Hartnoll 1988).

8.3.1 Mechanical Support and Locomotion

Like their aquatic relatives, amphibious and terrestrial crustaceans obtain all necessary mechanical support for both their internal organs and external activities from their hardened exoskeletons. The basic architecture of the exoskeleton and the functional morphology of virtually all external movements, including locomotory movements, appear to be essentially the same within each major group of crustaceans independent of whether they live and move about in water or on the land. Whatever differences result from the loss of buoyant support in subaerial habitats seem likely to be fully compensated for by variations in relevant physical properties (mechanical strength, flexibility, density, thickness, etc.) of exoskeletal parts. Variations in amounts of calcification are almost certainly one of the major bases for variations in exoskeletal properties. The anterior parts of the exoskeletons of many semiterrestrial crustaceans that dig or burrow are more heavily calcified than are other parts, apparently to aid in abrasion resistance (Greenaway 1985).

Crustacean exoskeletons in general are complex composite structures made of the same anatomical elements (thin epicuticle, procuticle of variable thickness, and a single-cell layer of epidermis) as are the integuments of other arthropods (Neville 1984). Calcification

TABLE 8.2
*Structural and Mechanical Features of Amphibious
and Terrestrial Amphipods and Isopods*

Preadaptations	Newer Developments
AMPHIPODS:	
Lateral compression of body	
Thin, light weight integument	Exoskeleton relatively fragile, smooth, no outer waxy layer
Pleopods used for swimming, respiration, some feeding	Pleopods often reduced (lost in some species), occasionally used for respiration and locomotion
Strong pereopods for burrowing and locomotion	Gills on pereopods variably developed and modified for use in air
ISOPODS:	
Dorsoventral compression	
Pleopods used for respiration (both rami modified as gills) and for swimming	Pleopods modified in various ways in different groups: exopodites may be modified as protective coverings in some; capillary grooves in surfaces facilitate water transfers from uropods to gills in others; many with pseudotracheae possibly used for respiration
Pereopods strong for burrowing and locomotion	

SOURCES: Powers and Bliss 1983, Friend and Richardson 1986, Spicer et al. 1987, Warburg 1987.

occurs in the procuticle, resulting in increased stiffness and hardness (Hillerton 1984; Stevenson 1985). Beyond these very general statements strikingly little is known about the exoskeletal biomechanics of crustaceans. This is clearly a potentially interesting and fruitful area for future investigations. There seems to be nothing quantitative known about the probable biomechanical differences that are likely to exist between the exoskeletons of aquatic species and those of amphibious or terrestrial forms.

The major structural and mechanical adjustments for subaerial life shown by amphibious and terrestrial crustaceans may be viewed as resulting from combinations of evolutionary preadaptations, which existed in aquatic representatives of the group in question, and newer modifications and developments that have evolved either during the transitions to land or subsequently. Table 8.2 summarizes these adaptations for terrestrial amphipods and isopods. The crayfishes show little change. Brachyuran and anomuran crabs show substantially greater within-group variability.

Studies of crustacean locomotion in the context of comparisons between water and air are few. Most are qualitatively descriptive (Spicer et al. 1987). There are a variety of adaptations for terrestrial movement among the *crabs* (Dunham and Gilchrist 1988). The basic repertoire of movements is as in aquatic crabs, with most walking or running sideways, but all are able to move in any direction. Hermit crabs on land typically walk in the forward direction. Straight-ahead locomotion is also primary for the many tree-climbing crabs (over 560 species described, belonging to five families). Different groups of tree-climbers vary in climbing patterns: some face upward while climbing and downward while descending, while others face downward at all times. Tree-climbing hermit crabs of the family Coenobitidae tend to climb by spiralling up the tree trunk, the direction of the spiral usually being in the direction of the aperture of the shell they are carrying. Shelled hermit crabs trying to descend facing downward often lose their grip and fall to the ground.

When walking on relatively level surfaces *brachyuran crabs* typically use only eight legs. During usual sideways movements the four trailing legs push the crab, while the leading legs are used for steering and raising the body over obstacles. The heavily clawed chelipeds are held off the ground and do not participate in locomotion. The neighboring legs on each side of the crab alternate in moving, and they also alternate with the contralateral legs. This results in a physically very stable, alternating tetrapod gait which is used up to moderate speeds.

The major locomotory differences between brachyurans in air as compared with those in water occur at higher movement speeds. Brachyurans in air, such as many grapsids and the ocypodid ghost crabs, can run and jump for substantial distances; the greater density and viscosity of water largely prevent such activities or greatly reduce their magnitude and velocity. Ghost crabs running in aerial environments change their gaits as they increase their speeds. At moderate speeds they become hexapods, using the first three pairs of walking legs and raising the fourth off the ground. At the highest speeds only the first two pairs are used and the crabs literally leap along (Herreid and Full 1988; Full and Weinstein 1992).

Anomuran hermit crabs on land move in essentially the same ways they do in water, since they normally carry with them relatively large and heavy protective snail shells. These loads make it almost impossible for a hermit crab to run very fast for any distance, or to jump. Walking in hermit crabs involves only two pairs of legs, since the posterior two pairs have been modified to hold and support the shell. Hermit crabs normally walk in the forward direction. The chelipeds are used for steering and support and for climbing over obstacles. The two locomotory pairs of walking legs provide thrust. Like many insects, many hermit crabs use an alternating tripod gait, which is also physically quite stable. Additional stability derives from the support given by the protective shell, which is usually dragged along behind (Herreid and Full 1988).

A detailed consideration of the physical determinants of the differences between aquatic and subaerial pedestrian (walking to running) locomotion is complex and beyond the scope of this discussion. Daniel and Webb (1987) provide an introductory summary of each of the major parameters. These parameters are as follows:

- Locomotion is basically a problem of momentum transfer between animals and their environments.
- Momentum balances for thrust and resistance show that densities and viscosities of ambient media are the principal properties affecting the transfer of momentum. Both differ substantially between air and water.
- Propulsion mechanisms can be described in terms of forces generated (i.e., drag, lift, acceleration reactions in fluids, and reactions against the ground).
- Performances of these mechanisms can be described in terms of speeds, accelerations, and costs of transport.
- Two major factors influencing the distribution and performance of propulsors in all environments are the nature of unsteadiness in the speeds of animals and the density ratios between the animals and their environments.

These physical and engineering considerations have a wide range of specific biological manifestations and consequences in each of the crustacean groups we discuss. In addition to the qualitative investigations of gaits, tree climbing, and jumping already described, a small number of quantitative studies have been made of two categories of such considerations—(1) neuromuscular motor patterns and limb movements and (2) metabolic costs of locomotion.

The first category is analyses of neuromuscular motor patterns and walking limb movements of decapods in and out of water. In this work the amphibious capacities of the subject animals have been used as the basis for varying the mechanical loads carried by the limbs—the different density ratios between the animals and their environments in water and air producing natural load variations. Two aspects of the results are relevant here. First, the detailed character of locomotory movements changed with varying loads in forward walking, unrestrained crayfish (*Procambarus clarkii*). At similar walking speeds, frequencies of unloaded (underwater) walking movements were 50 percent higher than frequencies of loaded (in air) movements. This frequency difference was due primarily to longer durations of power strokes of the legs in animals in air; the return strokes were constant in duration over the observed step frequencies in both environments. The major power muscles in the legs (anterior depressor muscles) were much less active in the unloaded (underwater) condition. The mechanical advantage of the power strokes was also increased under conditions of load (in air) by shifting the arc swung by each leg towards the rear, and by reducing base-to-tip distances of the leg (Grote 1981).

The second relevant result of the air-versus-water experiments pertained to sideways-walking shore crabs (*Carcinus maenas*). The results were similar to those from the crayfish. The overall electromyogram patterns were essentially the same for crabs moving in both water and air. The power strokes for leg movements on land lasted longer than for those in water, thus step periods for spontaneously moving crabs were longer in air. Power strokes in air involved the activation of additional motor neurons in locomotory muscles innervated by several motor neurons, presumably increasing the strengths of contractions by spatial summation. Maximal frequencies of motor nerve stimulation were also higher during power strokes in air, presumably further increasing strengths of contraction by temporal summation (Clarac et al. 1987).

These observations, although they involve only two species, each belonging to a different group of decapods, are probably indicative of patterns that are phylogenetically widespread among the decapods. Whether or not this is true, however, it seems highly probable that at least comparable and analogous changes in locomotory movements and their associated neuromuscular mechanisms and controls occur in all crustaceans that move between water and land (this is supported by the recent study of walking movements of *Pachygrapsus crassipes* in and out of water, presented in Hui 1992). Thus it seems safe to say that land transitions in these groups probably did not involve substantial changes in what might be called the architectural and hard-wired portions of locomotory neuromuscular motor and sensory systems but did involve significant adaptive changes in the programmable software components of these systems.

The second category of biological consequences of engineering considerations that has been quantitatively analyzed is that of the metabolic costs of locomotion in amphibious and terrestrial forms. Here again the literature is severely limited in scope. The only significant data set relating to the quantitative aspects of locomotory energetics is restricted to about ten species of *anomuran and brachyuran crabs* (Herreid and Full 1988; Full and Weinstein 1992). The small amount of activity-related information available for the other crustacean groups involves only routine metabolic rates (Powers and Bliss 1983; Wieser 1984).

Herreid and Full (1988) and Full and Weinstein (1992) thoroughly and carefully review the literature on land crabs. Four of their conclusions are relevant to our discussion. First, the basic features of land crab activity energetics in air are very similar to those of vertebrates. Second, a substantial range of variation in the patterns of metabolic responses to exercise is found in different species of crabs. A complex suite of alternate scenarios exists for possible patterns of adaptation to subaerial life by ancestral decapods. Third, very few data permit direct comparisons of the metabolic costs of comparable levels of activity in crabs in water and in air. Fourth, these few studies are not consistent internally and seem not to have taken adequate account in their experimental designs of the differences in buoyancy and drag of the two environments. This is evidenced by what must be termed a preliminary result, that the three species of brachyurans studied showed comparable costs of transport for walking in both environments. Since the crabs were almost ten times heavier in air than in water an implication of this result, if it is correct, is that reductions in metabolic rates due to lower loading in water were exactly offset by increased metabolic rates compensating for the greater drag associated with moving in water. Further work is needed.

Herreid and Full (1988) also point out that data are currently not available on the energy costs of almost all other aspects of the activities of semiterrestrial and terrestrial crustaceans aside from more or less continuous, rectilinear locomotion. Thus it is not possible to even begin to develop for most of these forms meaningful energy budgets that might shed some light on the relative energy costs of life in air as compared with water. A first step toward filling this information gap has been taken by Full and Weinstein (1992) in their work on intermittent activity in a ghost crab on land.

Finally, to complete this section we note that the entire discussion to this point has dealt with crustaceans during their intermolt stages as juveniles and adults. Almost nothing seems to be known about possible special features of eggs and very early developmental stages (not necessarily larvae, since many amphibious and terrestrial crustaceans have direct development of juveniles) in the context of cuticular properties, modes and energetics of locomotion, and other issues.

Molting, however, is a subject that has been, and remains, the focus of much research interest and activity. The basic processes, mechanisms, and controls of molting appear to be uniform across all groups of crustacea, including the amphibious and terrestrial forms (Greenaway 1985; Skinner 1985; Stevenson 1985). Molting in subaerial environments, however, imposes on crustaceans at least three important stresses that are less serious for aquatic forms or for semiterrestrial forms that either retreat into water or retire to wet or moist burrows to molt. These stresses are: the risk of desiccation, the need to acquire water for size increases, and the need to acquire calcium salts to harden the new integument (Hartnoll 1988).

Once again, the major group in which these matters have been studied in detail is the decapods, specifically the land crabs. Hartnoll (1988) reviews the literature, reaching several important conclusions. First, with respect to the water-related stresses, little is known about anomurans that do not carry shells with them. Anomurans that carry shells apparently carry water in them as well when molting approaches. Brachyuran land crabs have unusually well-developed and enlarged pericardial sacs (pouchlike organs found at the rear of the branchial chambers; they have cavities that communicate with the pericardial sinuses of the circulatory system). Foregut and midgut glands can also store water. All three structures are gradually filled with water (using several different mechanisms) during the premolt periods. This stored water is sufficient both to prevent dangerous desiccation before a new shell is hardened and to ensure that enough water is present to facilitate the concomitant increase in body size.

Hartnoll's second conclusion is that calcium salts are obtained both from the diet during intermolt stages and from reabsorption from the old integument during premolt stages. These salts are stored in various sites within the body, notably the gastroliths (whitish concretions that develop between the epidermis and the cuticle of the foregut; they, and other less widely occurring calcium salt storage structures, are not shed at ecdysis). These internal calcium stores may provide for hardening of the chelae and mouth parts. The crabs can then eat their old exoskeletons and recycle their calcium. Most, if not all, land crabs eat their molts. The stored salts are adequate for initial postmolt hardening of at least parts of the new integument. Salts for later hardening derive from the food.

There is, however, a cost associated with both these adaptations (this is Hartnoll's third conclusion). Molt increments in growth in land crabs appear to be relatively smaller than those in many aquatic crabs. The frequency of molting in land crabs is often negatively affected by seasonal water shortages. As a result, growth in size of land crabs is often slow. However, many of them seem relatively long-lived and often reach large sizes.

The molting cycle necessarily also has a fourth major set of interactions that must have substantially influenced the processes of terrestrial adaptation in all crustacean groups. The massive shifts in calcium salts that occur (the major salts are various mineral forms of calcium carbonate and calcium phosphate [Stevenson 1985]) interact strongly with: (1) the processes of extra- and intracellular acid-base regulation; (2) ionic transport mechanisms in the gut, gills, and excretory organs; and (3) respiratory gas exchanges (especially carbon dioxide). Amphibious decapods specifically compensate potential subaerial metabolic and respiratory acidosis by mobilizing exoskeletal calcium carbonate. This may be considered a preadaptation for buffering against acidotic episodes that are directly related to the molt cycle (Taylor et al. 1991). These and other related subjects are discussed in the next two sections.

8.3.2 Respiration, Circulation, and Metabolism

Comparative, often also ecological, respiratory physiology is now one of the most active research areas in animal physiology. Amphibious or terrestrial crustaceans, almost always decapod crabs, are often subjects of this work. Thus the literature relevant to these discussions is growing rapidly.

The major contrasts between water and air as respiratory media, also the design and regulatory contrasts between respiratory systems in aquatic and subaerial animals, have been thoroughly reviewed (Bicudo and Weibel 1987; Dejours 1987a, 1987b, 1988; Rahn and Dejours 1987; Truchot 1987, 1990; Cameron 1989). There is nothing new for us to add on these subjects.

Amphipods Table 8.2 summarizes some general morphological differences that are thought to have respiratory significance between the gills and pereopods of terrestrial as compared with aquatic amphipods. Little (1990:130) points out that in some semiterrestrial forms the gill areas are reduced in comparison with some aquatic species, and the gills are more solid. These variations may relate more to the humidities of the subaerial environments in which the amphipods are active than to respiratory needs.

There is a general correlation between degrees of terrestriality in various talitrid amphipods and their abilities to survive indefinitely submerged in water or in air and to respire effectively in either medium. Most supralittoral talitrids studied so far survived indefinitely in well aerated water, though a few died after fairly long periods. This was not the case for the single species of fully terrestrial amphipod studied (*Arcitalitrus dorrieni*—a landhopper), which drowned within a few hours. The abilities of different amphibious species to maintain constant (for given temperatures) rates of oxygen consumption in both environments were variable. The ratio of aerial to aquatic oxygen uptake ranged (for seven species) from 0.85 to 2.30; three of the seven showed ratios indistinguishable from 1.0. The bases for this variability are not clear, but may include: (1) variations in levels of locomotory activity by the animals within the respirometers during experiments (oxygen uptake rates were not measured continuously, and the animals were not observed regularly); (2) variations in relative amounts of general cuticular oxygen uptake as compared with branchial uptake; and (3) interspecific variations in relative sizes of gills (Spicer and Taylor 1987; Spicer et al. 1987).

There is also evidence that species of beachfleas and landhoppers that are more fully terrestrial can maintain constant rates of aerial oxygen consumption for longer periods at higher temperatures than can less fully adapted species. Further, those species that are more fully terrestrial show lower thermal sensitivities (Q_{10} values) for aerial oxygen consumption than do less fully adapted forms. These differences are interpreted as evidence that the more fully adapted terrestrial forms can better tolerate the potentially greater thermal variability of terrestrial habitats. Note, however, that these data derive from work with a small number of species from only one geographic region—the British Isles, though one species studied had been introduced from the southern hemisphere (Spicer et al. 1987; Spicer and Taylor 1987).

The only amphipod species yet studied with respect to the oxygen transporting properties of the hemolymph are two semiterrestrial beachfleas. Thus there is no way of determining whether or not the properties found relate to degrees of terrestriality or to other environmental factors important to these particular forms (e.g., tolerance for hypoxia). Spicer and colleagues (1987) speculate about these matters in relation to the hemocyanin found in these organisms having a higher oxygen affinity and a more moderate Bohr shift than the hemocyanins in some decapod crustaceans.

Isopods Respiratory structures and physiology among the amphibious and terrestrial isopods are varied and not clearly separable from those of aquatic forms (Wieser 1984; Warburg 1987). Little points out that isopods carry out much of their respiratory gas exchange via their pleopods, the rest via the ventral surfaces of their abdomens (1990:221; see also table 8.2). There is some variation between species in the relative amounts of exchange occurring via the two routes, but the total rates of exchange are less affected by the routes used than they are by environmental conditions such as aerial humidity. The pseudotracheae present in many forms of isopods are considered primarily adaptations for gas exchange in desiccating environments.

Ecophysiological studies of water-land transitions by the isopods have focused almost exclusively on water and temperature relations. There seems to be nothing substantial known about respiratory properties in these contexts. The application of modern approaches would be likely to produce an interesting array of new information.

Decapods Respiratory adaptations for subaerial environments among amphibious and terrestrial decapods have been studied extensively. The crayfishes are the least studied group.

Many kinds of *crayfishes* are amphibious burrowers. They leave their usual freshwater homes, most often nocturnally, to forage on the adjacent land or to move from one body of water to another. Many species, either in response to drought conditions or apparently to position themselves closer to favorable feeding areas, dig large and complex burrows with their surface openings often quite distant from the nearest surface water. These burrows usually extend down to the ground water table (Hobbs 1981; McMahon and Wilkes 1983; Horwitz and Richardson 1986; Atkinson and Taylor 1988; Hart and Clark 1989).

Under laboratory conditions, crayfishes of the types just described will spontaneously leave normally oxygenated (normoxic) water or they will "emerge" in response to hypoxia in their water. Aspects of emergence-related adjustments in respiratory physiology of two species have been studied in some detail: the North American *Orconectes rusticus* (McMahon and Wilkes 1983) and the most thoroughly studied form, the European *Austropotamobius pallipes* (much relevant work is reviewed in Taylor et al. 1991 and Wheatly and Henry 1992).

Both species show variable emergence behaviors, depending on specific environmental conditions. Spontaneous emergences from normoxic water are often brief (one to two minutes duration), but under simulated burrow conditions (moderate temperatures, moderate to high aerial humidities) emergences can last many hours. The animals can survive in air, remaining fairly active, for at least forty-eight hours. Emergences from hypoxic, or hypoxic and hypercapnic (high carbon dioxide), waters usually last five to ten minutes and are often partial. The animals frequently emerge just far enough to expose one or both inhalant branchial apertures—thus allowing air to enter the branchial chambers—and then they will resubmerge, releasing bubbles from the exhalant apertures for some time afterward.

Physiological responses to these different conditions are complex and vary between situations. In particular, there are large differences between responses to brief air exposures as compared with long-term exposures, and between spontaneous as compared with forced air dives. There are also significant differences in behavior between the two species, *O. rusticus* tending to emerge partially and periodically while *A. pallipes* tends to emerge completely for longer times. These behavioral differences are reflected in detailed physiological differences as well.

Overall, the fact most relevant to this discussion is that these two amphibious crayfishes show some similarities to vertebrate breath-hold divers in their short-term responses: rates of oxygen consumption initially drop rapidly and changes in both breathing and heart rates occur, all associated with a combined metabolic and respiratory acidosis. With longer exposures these initial changes are fully compensated for over extended periods (one to two days). The compensations involved include: (1) partial drying of the gill surfaces, allowing the initially somewhat collapsed gills to separate from each other and resume gas exchange; (2) mobilization of carbonate from the exoskeleton to offset the acidosis resulting from both initial anaerobic production of lactate and reduced carbon dioxide losses; and (3) additional lactate removal by a combination of excretion, tissue sequestration, and metabolism.

After about forty-eight hours conditions again begin to deteriorate, probably because further gill desiccation interferes with gas exchange and the absence of contact with liquid water interferes with necessary ion exchanges and waste excretion. Crayfish returning to water after twenty-four hours in air initially hyperventilate; but oxygen consumption, ventilation and heart rates, and all affected blood parameters return to settled submerged levels within eight hours.

Taylor et al. (1991) consider this suite of interactive responses to air exposure in crayfishes as an example of preadaptation. The various processes involved possibly derived evolutionarily from adaptations of submerged animals for hypoxia, acid water conditions, and heavy exercise.

The changes in blood chemistry associated with air exposure necessarily have major impacts on the hemolymph, notably, the acid-base status and the oxygen-carrying capacity. The latter change occurs primarily via modulation of the oxygen-binding properties of hemocyanin. *A. pallipes* is the only crayfish in which these processes have been studied. In this species the short term, acidotic drops in blood pH are compensated by the sequestration of some hydrogen ions in the muscles and by mobilization of a buffering base (exoskeletal calcium carbonate), the latter also producing an increase in calcium ion concentrations in the blood. The hemolymph *in vivo* shows a high oxygen affinity at lower pHs despite a normal acid-induced Bohr shift to the right of the hemocyanin oxygen-loading curve. The Bohr shift results from specific allosteric effects of both calcium and lactate ions. Thus the degree of oxygen saturation of the hemocyanin is kept high even though oxygen uptake rates by the gills are initially reduced. After the normal pH is restored in the blood, the Bohr shift of the hemocyanin is eliminated, by reversal of the allosteric effects of calcium and lactate. Hemocyanin oxygen affinity is further increased as a result, and normal rates of oxygen transport are restored (Taylor et al. 1991).

We have devoted substantial attention to the physiological details of what is known about only a few species of crayfishes for several reasons. First, the lack of species diversity in the available data sets does make it impossible to say to what degree the results are general, even for the crayfishes, but it also simplifies and clarifies what is already a complex series of events and interactions. Second, the relative clarity of the crayfish story makes these data easier to use as a basic model for discussing the much better documented, more varied, and (in some respects) more complex situations found in the land crabs. Third, important gaps in our understanding of crustacean respiratory, circulatory, and metabolic physiology, in the context of water-land transitions, are highlighted by the relative completeness of the picture for a few crayfishes. As examples, contrast the sophistication and completeness of our understanding of air emersion effects in *A. pallipes* with the partial and fragmented pictures we have of amphipods and isopods, or the complete lack of information that exists with respect to the amphibious burrowing marine thalassinid, *Thalassina anomala*, in the Far East, and the tree climbing marine shrimps, family Hippolytidae, of mangrove areas in the southern Caribbean and Malaysia (Powers and Bliss 1983:285–86).

Anomuran and brachyuran crabs Present knowledge of the biochemical, physiological, behavioral, and ecological diversity of the amphibious and terrestrial anomuran and brachyuran crabs, even in the limited context of this section, greatly exceeds that of all of the other groups discussed so far. Since many of these crabs, at least as adults, are fairly large, often common, and are relatively easily maintained in laboratories, they have attracted the research interest of many people. The relevant literature is large and continually expanding, and it has been reviewed recently several times (Burggren 1992; Burggren and McMahon 1988; Burnett 1988; deFur 1988; Greenaway and Farrelly 1990; McMahon and Burggren 1988; Taylor et al. 1991; Truchot 1987, 1990; Wheatly and Henry 1992). In the rest of this section we summarize eight of the major conclusions deriving from work on land crabs.

1. The anomuran land crabs include among their numbers (from three families) representatives showing grades of terrestrial adaptation from T-1 to T-4. (See table 8.1 for a definition of each grade; T-1 is the least terrestrially adapted.) The brachyuran land crabs (from at least 17 families) show grades from T-1 to T-5 (Hartnoll 1988). The principal organs of external respiration in all species are the gills and the linings of the branchial chambers. A few genera of ocypodid crabs (e.g., *Dotilla* and *Scopimera*) have evolved accessory cuticular surfaces for respiratory exchange that are located in unusual places— on the walking legs or the ventral thoracic sternal plates. The general structural trend is for the gills to become reduced in both overall size and surface area as the degree of terrestriality increases, though gills are retained in even the most fully terrestrial forms. Associated with these gill changes are enlargements of the branchial chambers. In those forms with greatly expanded chambers the branchiostegal membranes used for respiration are relatively smooth. Forms with less expanded chambers usually show specialized modifications of various parts of these membranes that result in increased surface areas of those parts. The modified portions act as accessory respiratory organs. The soldier crabs of Australia (genus *Mictyris*) have particularly highly evolved structures in this context (Maitland and Maitland 1992).

2. This morphological diversity is accompanied by lesser functional diversity in terms of ventilatory mechanisms. The bilateral scaphognathite pumps can, in most species, pump either water or air and can do so equally effectively. The higher oxygen content of air often permits lower pumping rates in aerial environments, but the maximum ventilatory performance is generally comparable in both

air and water. A relatively few species of land crabs, both anomurans and brachyurans, also use other types of ventilatory pumps. Some make lateral movements of the abdominal mass, which acts as a piston driving air in and out of the branchial chambers. A few species have been found to use muscles to lift or lower the carapace, which may have respiratory effects.

3. Crabs vary considerably in the degree to which they carry water with them in their branchial chambers while they are out of water. The less fully terrestrial forms that do not carry water must revisit water more frequently to keep their respiratory surfaces moist. Water retained in the branchial chambers apparently serves several important functions: it helps prevent collapse of gill structures, thus maintaining larger surface areas for exchanges of respiratory gases; it serves as an exchange medium for ions involved in carbon dioxide elimination and acid-base balance; and it helps maintain levels of body hydration and reduces rates of evaporative water losses from the bodies of the crabs.

4. Land crab metabolic rates, both rates of oxygen uptake and carbon dioxide elimination, vary considerably. They are affected by the same factors that affect metabolic rates in most ectothermic animals (e.g., temperature—both acutely and with chronic acclimations—season, and size) and also by several more specific factors (e.g., stage in the molting cycle and lengths of time out of contact with liquid water in the more amphibious forms). Given all these factors it is not surprising that metabolic variability persists, even for the relatively few species that have been studied in detail comparable to that described in the previous section for one species of crayfish. It is also the case, as pointed out earlier, that few data are available comparing metabolic rates within single species of crabs, in and out of water, at comparable levels of locomotory activity. Thus there is no present possibility of deriving general principles as to metabolic rates from the literature.

5. With respect to both extracellular acid-base balance and the effects of air emersion on oxygen and carbon dioxide transport properties of the hemolymph, land crabs generally appear quite similar to the crayfish, *A. pallipes*, described earlier (Wheatly and Henry 1992). Potential metabolic and respiratory acidosis associated with initial collapses of gill structures are compensated by mobilization of calcium carbonate from the exoskeleton. In some species increased

hemolymph levels of calcium ion and lactate cause short-term bene-
ficial increases in the oxygen affinity of hemocyanin; these increases
help restore oxygen transport and aerobic metabolism. Adjustments
similar to those described for the crayfish occur in the amphibious
species of crabs after longer exposures in air. In the more terrestrial
species the oxygen partial pressures of hemolymph are high and
carbon dioxide partial pressures are low. The former is generated
primarily by oxygen exchanges across the branchiostegal membranes
(the "lungs" of land crabs), the latter by carbon dioxide exchanges
across the gill surfaces. If water is absent in the branchial chambers,
carbon dioxide also exchanges across the branchiostegal membranes
(Henry 1991).

6. Respiratory ventilation and hemolymph circulation are closely
coupled in aquatic crabs and, apparently, in land crabs submerged in
water. Countercurrent exchanges of respiratory gases are important
features of aquatic respiration in decapods. In land crabs not maintaining significant volumes of water in their branchial chambers
while on land, these exchanges necessarily cease and are replaced by
diffusional exchanges that are generally less precisely oriented. It is
possible, however, that in a few species, such as the anomuran
coconut crab, *Birgus latro*, a gas exchange system closer to the multi-
capillary system found in the lungs of birds may operate (Greenaway
et al. 1988). Associated with these changes in gas exchange processes
are changes in the coupling of ventilatory and cardiac activities (Wilkens and Young 1992).

7. In a few species of land crabs there is evidence of some capacity
to actively direct and control patterns of hemolymph flow in response to changes in rates of respiratory gas exchanges across the
gills as compared to exchanges in the branchiostegal membranes.
The gills are more effectively perfused when the crab is immersed in
water, the branchiostegal membranes when the crab is in air. There
is also good reason to expect, though no direct evidence is apparently
yet available, that land crabs make a variety of cardiovascular adjustments to the effects of gravity, associated with the loss of buoyancy
of water, when they emerge onto land.

8. To summarize, we may say that the extant land crabs demonstrate a wide variety of respiratory, circulatory, and metabolic mechanisms as parts of their overall adaptations for the water-land transi-

tions each organism makes many times over in the course of its life. Most are based on morphological and physiological capacities preexisting (and hence preadapted) in the aquatic crabs, in which they are used for other purposes.

8.3.3 Thermal, Water, and Solute Relations

The subject areas of this section have for many years formed the major bases for both comparative physiological and evolutionarily oriented studies of the terrestrial invasions made by crustaceans. Little (1983, 1990) relies on thermal, water, and solute relations in his constructions of scenarios for both the possible sites and processes involved in the land transition. Since so much attention has already been given to these subjects, we restrict ourselves here to only what we consider the most important and relevant results. Once again there are a number of recent reviews: Powers and Bliss 1983; Takeda 1984; Wieser 1984; Spicer et al. 1987; Atkinson and Taylor 1988; Greenaway 1988; Macintosh 1988; Truchot 1990. Also once again the largest volumes of data giving the most detailed pictures derive from the *brachyuran land crabs*.

Amphibious and terrestrial crustacea moving between aquatic and terrestrial environments, or living entirely in terrestrial environments, must deal with the standard set of environmental stresses relating to temperature, water, and solutes. One stress is that ranges of temperatures encountered in subaerial environments are likely to be larger and are more likely to exceed physiological tolerance limits of animals than are temperature ranges in aquatic environments. Desiccation, too, is a near universal, continuing hazard in subaerial environments, except for such microhabitats as deeper cracks in rocks or burrows in soils. Open and exposed subaerial environments often produce strong interactions between thermal and water relations. For example, higher ambient temperatures and movements of ambient air are both likely to increase rates of evaporative water loss, which in turn lowers body temperatures. Higher metabolic rates associated with higher body temperatures in subaerial situations are likely to increase rates of excretory losses of solutes, requiring additional efforts to compensate for these losses.

This list could be expanded, but the main point is clear: occupation of terrestrial habitats requires a complex, interacting suite of adjustments by crustaceans with respect to all three variables—thermal, water, and solutes. Some of the most widespread features of this suite of adjustments are:

- Behavioral mechanisms of many kinds are used to moderate or eliminate the most severe stresses.
- Behavioral thermoregulation and behavioral regulation of states of body hydration are two common patterns.
- Nocturnality is widespread among more completely terrestrial forms and rhythmic patterns of activity keyed to tidal cycles are near universal in intertidal and littoral forms.
- Many of the more completely terrestrial forms show somewhat greater abilities to tolerate desiccation than do more amphibious forms, but usually not by a large margin.

There are few other consistent physiological differences relating to these three variables that differentiate between crustaceans belonging to the different grades of terrestriality listed in table 8.1.

Amphipods The thermal physiology of amphibious and terrestrial amphipods is essentially unstudied. One species of beachflea and one landhopper, both living in temperate climates, have been shown to be tolerant of a wide range of temperatures (with lower and higher lethal temperatures near −3°C and 37°C, respectively). The landhopper demonstrated seasonal thermal acclimation, in that summer upper lethal temperatures were a few degrees higher than winter upper lethal temperatures. Amphipods under field conditions universally use behavior (hiding under objects, burrowing) to avoid direct exposures to extreme temperatures. It is not known if there are differences in these respects between aquatic as compared with more terrestrial forms (Spicer et al. 1987).

    Among the amphipods ability to survive continuous aerial exposure without access to liquid water is limited, even among the most terrestrially adapted landhoppers that have been studied. In experimental situations maintenance of levels of aerial relative humidity (RH) near 100% at moderate, ecologically relevant temperatures is not easy to do, so published data on amphipod survival times at different humidities are not altogether reliable. Many members of the family Talitridae can survive indefinitely in air at or near 100% RH.

However, even the most terrestrial landhoppers studied have survived no longer than two to three days at 95% RH at 15°C. At lower humidities survival times were much shorter. Several species of beachfleas and sandhoppers were found to survive 95% RH for only ten to eighteen hours. At lower humidities survival times became so short that differences between species were no longer detectable (Spicer et al. 1987).

Limited data from a few, taxonomically widely scattered species of amphipods indicate that the group has no active ability to regulate rates of evaporative water loss. The primary correlation is with body size, thus presumably surface-to-volume ratio. Freshly dead animals evaporate water at the same rates, under given conditions, as living animals. Evaporative water losses greater than about 25% of body mass are usually lethal. These losses appear to occur across the general body surface. It is possible that at least some species secrete an epicuticular wax layer, but there is currently no good evidence that this has any significant effect on rates of water loss. Behavioral grooming, keeping the body surfaces free of potentially abrasive particles, may be important in reducing desiccation hazards in more terrestrial forms (Spicer et al. 1987). Finally, there is no evidence that any amphipods can actively absorb water vapor from the air, as has recently been found to be possible for at least a few fully terrestrial isopods (Wright and Machin 1990).

Observations of only a few species of amphibious and terrestrial amphipods revealed that the group appears to be quite variable in its osmoregulatory capacities. Beachfleas and sandhoppers are capable of living in a wide range of salinities (euryhaline), and they are fairly strong osmoregulators—hyper-regulating in dilute media and hypo-regulating in more saline media. Landhoppers are almost equally euryhaline, but are primarily osmoconformers at levels significantly above ambient osmotic concentrations. Within their naturally selected environments they regulate the osmotic concentrations of their body fluids at fairly high, but variable levels (400–840 mOsm/l). It is not clear that these properties are notably different from the osmoregulatory abilities of completely aquatic (either marine or freshwater) taxa of amphipods that are phylogenetically close to the species studied (Spicer et al. 1987; Morritt 1988).

Transitions from water to land by amphipods necessarily also had important implications for their ionic regulatory capacities and mechanisms. The problems are similar to those described for the molluscs in the previous chapter, and they can also serve as a paradigm for our discussions of the isopods and decapods.

As the different evolutionary lines of amphipods became more terrestrial, a corollary of their greater independence from liquid water was a reduction in their access to salts contained in that water. The severity of this reduction varied, of course, in relation to both the nature of the environments in which the transitions occurred and the physical distances that separated them from their ancestral water bodies. The most important general effects, for all groups, were reduced availabilities of sodium and chloride (the primary ions contributing to hemolymph osmotic concentrations), calcium (central to exoskeletal hardening after molting), and copper (essential for hemocyanin synthesis). The available sources of these materials also changed—to liquid water that might be encountered and perhaps consumed and to their food, whether plant or animal. Cutaneous and/or branchial uptake pathways and mechanisms were largely inactivated.

Stating these problems and the broad options available for solving them has turned out, in the context of the amphipods, to be much easier to do than documenting the actual strategies employed by various members of the group. Very few species have been studied thoroughly, almost none in all important respects. The available data base is too limited to permit firm statements differentiating which variabilities in properties occur within groups as well as between groups (e.g., aquatic vs. semiterrestrial or terrestrial). No conclusions can currently be reached with respect to sodium and chloride relations. Calcium relations apparently vary substantially between different evolutionary lines, some lineages storing considerable calcium for later use in hardening the exoskeleton (storage is in calcareous concretions in the posterior caeca of the midgut), others not. Copper metabolism is also variable and, on present knowledge, weakly correlated with environmental situations. Thus further work is needed to clarify all these matters (Spicer et al. 1987).

Isopods As mentioned earlier, the thermal, water, and solute relations of fully terrestrial isopods have been subjects of substantial research interest (Wieser 1984; Warburg 1987; Wright and Machin 1990). However, only one species of amphibious isopod, *Ligia ocean-*

ica, has been studied in these respects. The main results have been reviewed by the person who performed the original work (Edney 1968).

L. oceanica is a widely distributed, littoral (rocky shoreline) isopod that is primarily active nocturnally. It spends much of the daytime, especially during warm seasons in the warmest parts of the day, under larger rocks or in deep crevices where atmospheric relative humidities (RH) are near 100% and temperatures are usually lower than they are on the exposed surfaces of the rocks at the same times. In the laboratory, at 100% RH the upper lethal temperatures (ULT) for these animals were near 33°C for one-hour exposures and near 27°C for 24-hour exposures. At 50% RH the upper lethal temperatures were 34°C and 4°C, respectively. In dry air (0% RH) ULT was 39°C for one-hour exposure time; no animals survived this condition for 24 hours.

Field observations have helped in the interpretation of these results. In the field, when temperatures in refugia neared the short term ULT, the animals emerged onto the exposed rock surfaces—despite the fact that these surfaces were even warmer than where they had been. They remained in the open for the relatively short time periods until temperatures started dropping again as the sun moved lower in the sky. Measurements of their body temperatures, both in the refugia and during the periods on the surface, showed that they were isothermal with their surroundings in the former situations, but substantially cooler (up to 8°C cooler) out in the open. Thus the indications are that this species, in which the adult is physically quite large for an isopod, uses a combination of behavior, possibly some convective cooling, and evaporative cooling to survive critical periods of high temperature exposure in nature. The laboratory results for one-hour exposures are consonant with this interpretation—the highest ULT, at 0% RH, probably resulting from the highest rates of evaporative water loss producing the lowest body temperatures. The deaths induced in the laboratory by 24-hour exposures are most likely the result of desiccation effects, the animals being unable to tolerate the massive water losses associated with either of the low RHs.

Comparable studies of several fully terrestrial isopods showed a variety of response patterns. To date no one appears to have made the effort necessary to sort out the multiple factors that are involved in producing these different patterns. The recent demonstration that at least a few species of fully terrestrial isopods can obtain water directly from vapor in the air complicates the analysis (Wright and Machin 1990). It is certainly desirable to obtain further information in these areas, using modern methods, for other amphibious species of isopods.

Edney (1968) considered four other features of the amphibious *Ligia oceanica* as supportive of their terrestrial adaptation. First, the isopod's relatively large adult size (body mass near one gram) implies a lower surface/volume ratio than obtains for many fully terrestrial species of isopods, most of which are significantly smaller. As a result, even though there is evidence that the cuticle of *Ligia* is more permeable to evaporative water losses, per unit area, than the cuticles of the fully terrestrial forms, under given conditions of exposure to unsaturated air adult *Ligia* survived longer than adults of the terrestrial forms. Thus larger adult body size may be part of the overall adaptation of the species.

Second, and as a result of its higher general cuticular water permeability, *Ligia* loses proportionately less water across its relatively simple, gill-like pleopods (swimming legs) than do terrestrial isopods having pseudotracheae as parts of their more complex pleopods. The absolute rates of evaporative water loss from *Ligia*'s pleopods are, for comparably sized animals, much higher than those rates for the terrestrial forms tested. Thus two important developments in land invasions by isopods probably were the evolution of more water-tight cuticles plus the evolution of pleopodal morphologies that permit fairly efficient respiratory gas exchanges while limiting rates and amounts of respiratory water losses associated with those gas exchanges.

Third, and as might be expected from its relatively high cuticular water permeability, *Ligia* is tolerant of substantial variations in hemolymph osmotic concentrations resulting from desiccation. Animals survive hemolymph concentrations up to 75% higher than average levels in normally hydrated individuals. No data are available permitting a clear comparison of this ability between *Ligia* and any fully terrestrial isopod (Warburg 1987).

Fourth and finally, whether or not there is any substantial relationship between osmoregulatory abilities in water and terrestrial success is not clear. However, *Ligia* is amphibious and, in water, relatively euryhaline (it can tolerate salinities from near 25–100% of sea water). It is a good osmoregulator over the range 50–100% sea water, but gradually loses regulatory capacities in more dilute environments. Even so, at lower salinities it is a better osmoregulator than at least one species of intertidal aquatic isopod (*Idotea granulosa*).

In the context of these results it is interesting to note that the only data available on the distribution of water within the bodies of isopods relate to fully terrestrial forms. In the species that have been studied over 50% of the body water is in the cuticle, the major internal organs contain up to 20%, and the rest is in the hemolymph. These proportions are highly variable, however, among different species, the sexes within species, and seasonally (Warburg 1987). Thus no generalizations are now possible with respect to how a terrestrial mode of life may be facilitated by changes in the distribution of water within the isopod body.

A final item particularly important to the water relations of both amphibious and terrestrial isopods is the widespread occurrence of the aggregation phenomenon. Strong gregariousness leading to the frequent formation of aggregations of large numbers of individuals, most often in partially or well sheltered locations, is a characteristic feature of the behavior of all known terrestrial isopods. A propensity for aggregation occurs throughout the life cycle and is regulated by the occurrence in the feces of an aggregation pheromone. Takeda (1984) reviewed what was known to that date about these substances and the behaviors they induce. His own work included two species of sea lice, *Ligia exotica* and *Ligidium japonicum*.

Aggregation tendencies in these two species were low when the animals were normally hydrated and in environments with higher ("optimal") humidities. In lower, desiccating humidities aggregation activity was intense; at supraoptimal humidities it was unstable. Sea lice in aggregations were substantially more resistant to dehydration than were nonaggregated animals and survival times were longer. Rates of evaporative water loss were reduced by aggregation, as were metabolic rates. These effects were quantitatively largest in the amphibious *Ligia*, as compared with two species of fully terrestrial isopods similarly studied. A correlation is that the cuticle is thinnest in *Ligia*.

Takeda (1984) concluded that the aggregation phenomenon is an important component of the overall adaptation of terrestrial isopods to life on land, primarily as a way of limiting evaporative water losses. It also has several secondary and tertiary effects that could be important: the reductions in metabolic rate mentioned, which in turn could reduce needs for foraging; a stimulation of rates of body growth, resulting in the more rapid achievement of the benefits of increased body size, such as lower surface-to-volume ratios; reductions of individual risks of predation; and facilitation of reproduction.

Decapods The general considerations outlined at the beginning of this section (pertaining to thermal, water, and solute relations) all apply strongly to the amphibious and terrestrial decapods. The literature relevant to these properties of decapods is so large and diverse, and has recently been so well reviewed (Burggren and McMahon 1988; Wolcott 1992), that we restrict our remarks to a few aspects we wish to emphasize.

First, thermal, water and solute relations of semiterrestrial crayfishes in contexts relevant to their amphibious behaviors have not been studied. Second, it is striking that land crabs are almost entirely tropical. A few species are distributed in subtropical and warm temperate areas, but none occur in cool or cold environments. As Hartnoll (1988) points out, one's intuition would be that cooler temperatures would reduce the hazards of desiccation, but it appears that other factors are more important.

Third, there is a great deal of variation between tropical and nontropical species in patterns of behavioral responses to thermal environmental conditions and the associated effects on water and solute relations. Both sets of species show a range of diurnal activity patterns during warm or hot periods, but the nontropical species superimpose on the diurnal responses behaviors that evidence strong seasonal cyclicities. Notably, during cold periods in winter the nontropical species typically become inactive and remain in their burrows. The nontropical crabs are also likely to have lower thermal limits for such things as locomotory activity, molting, and reproduction than do tropical forms (Powers and Bliss 1983).

Fourth and finally, many different adaptive scenarios are used by different groups of amphibious and terrestrial crabs with respect to water and solute regulation and metabolism (Greenaway 1988; Wolcott 1992). The osmotic and ionic regulatory abilities of different species range from osmotic and, to some degree, ionic conformity to very effective hyper- or hypoosmotic regulation with associated strong ionic regulation. This diversity is considered to reflect, at least in part, the evolutionarily ancestral habitats from which the different lineages of decapods made their invasions of the land (marine, brackish water, and freshwater origins are all indicated). The physiological bases for the diversity in methods of water and solute regulation and metabolism include a wide range of different mechanisms and processes. To obtain water different species may use: osmotic uptake during immersion in liquid water, drinking, uptake of soil water (based on several different methods), ingestion of preformed water in the food, production of metabolic water, and uptake of water condensed on their body surfaces. On the other side of the ledger all species necessarily lose water by evaporation (specific sites of evaporative loss vary somewhat), in their urine, and with the feces. Ionic exchange pathways and processes in land crabs immersed in water are the same as those of aquatic crabs. Ionic exchanges in crabs out of water necessarily occur via the gut and antennal organs. Substantial amounts of recycling of both water and solutes from urine occurs across the gills of a number of more terrestrial species (Greenaway et al. 1990; Morris et al. 1991; Wolcott 1991; Wolcott and Wolcott 1991). Molting out of water creates special problems, which are also dealt with in different ways by different groups.

8.3.4 Nitrogenous Metabolism and Excretion

Aquatic crustaceans generally are ammonotelic, meaning that ammonia is the principal excretary form for waste nitrogen. Amphibious and terrestrial crustaceans retain this characteristic to varying degrees. The detailed nature of waste nitrogen metabolism and excretion is poorly known in many groups. Very few species have been carefully investigated (Regnault 1987).

The few talitrid amphipods that have been studied are ammonotelic. Excretion of gaseous ammonia into the air occurs in at least one species. Waste nitrogen excretion in talitrids is thought to occur primarily via the digestive tract; the antennal organs are greatly reduced in size. Fully terrestrial talitrids show no signs of using uric acid as an important form of waste nitrogen excretion (Spicer et al. 1987).

Nothing is known about waste nitrogen metabolism and excretion in amphibious *isopods*. Fully terrestrial isopods are primarily ammonotelic, but small amounts of urea and uric acid are also produced.

The ammonia produced is released primarily as ammonia gas. The maxillary nephridia release bursts of urine into the ventral water duct systems of the animals. As the urine flows along these systems it becomes more alkaline, which leads to the volatilization of the ammonia. The interrelationships between rates of waste nitrogen excretion and rates of water loss, also diurnal and seasonal variations in waste nitrogen excretion rates, have been studied in several species of fully terrestrial isopods (Wieser 1984). The most arid-adapted species, however, remain unstudied (Warburg 1987).

Waste nitrogen excretion in the amphibious *crayfishes* in and out of water is unstudied; our assumption is that they are primarily ammonotelic. Surprisingly few species of *land crabs* have been studied in this respect, but those few have been carefully selected by the researchers involved. As a result, considerable diversity in strategies for dealing with waste nitrogen has been discovered.

The single anomuran studied so far, the almost completely terrestrial robber or coconut crab (*Birgus latro*), is uricotelic, meaning it uses uric acid and the salts of uric acid as its principal nitrogenous waste (Greenaway and Morris 1989). Among the brachyurans several species of amphibious gecarcinids have been studied. Most of these accumulate in their hemocoels substantial deposits of uric acid and urate salts, which apparently continue to build up as long as the crabs live. Under laboratory conditions these deposits can amount to as much as 55% of the total dry mass of a crab (Wolcott 1991). Actual nitrogen excretion from these species is in two forms of ammonia: ammonium ion or gaseous ammonia. The sole species shown to date to produce ammonia gas (*Geograpsus grayi*) releases this gas sporadically, from a still unidentified source thought to be the gills, in amounts as high as 80% of total nitrogen excretion. The remainder (a very small percentage of the total) is eliminated through the feces as ammonium ion and urea(Greenaway and Nakamura 1991). The other species, which use ammonium as their primary excretory form (up to about 70% of total nitrogen excretion), add this ion to the almost nitrogen-free urine produced by their antennal organs while the urine is being modified during contact with the gills in the branchial chambers prior to release of the excretory fluid. In these species as well nitrogen excretion via the feces represents a small fraction of total excretion (Greenaway and Nakamura 1991; Wolcott 1991).

8.3.5 Sensory Biology

There are many ways in which the sensory biology of crustaceans out of water must differ from the sensory biology of crustaceans in water. The multiple differences in the visual and auditory environments of the two media must have selected for substantial adaptations by at least some amphibious or terrestrial forms in such processes as object location, image formation, and direction finding. Aquatic crustaceans may rely more on olfaction than do terrestrial forms, while the latter may have better developed senses of taste.

It is striking that specific evidence is lacking for sensory differences that can be directly related to the water-land transitions in the various groups. An almost total absence of relevant comparative data seems the basic cause of this situation. For example, Burggren and McMahon (1988) and Rittschof (1992) mention these topics only briefly. Nevertheless, it is clear that the sensory structures of amphibious and terrestrial crustaceans are usually not substantially different from those of aquatic crustaceans. Steinbrecht (1984) and Land and Fernald (1992), however, have found in the *decapod land crabs* some exceptions to this statement. The general impression of uniformity may thus simply be one more artifact of our ignorance of actual levels of diversity.

Suggestive leads exist. Spicer et al. (1987) point out that the first antennae of talitrid *amphipods* are reduced in size as compared with the first antennae of aquatic amphipods. This is associated with a reduction in the sizes of the olfactory lobes of the brain—a correlation in line with the general prediction made earlier in this section. The situation may well be more complex among the *isopods*. Several types of completely terrestrial isopods are known to have well developed powers of chemical discrimination, apparently based primarily on olfaction (Wieser 1984). The sensory setae on the large antennae of isopods appear to be the sites of reception of the pheromones that are the basis for the aggregation phenomenon in these animals. These setae are less well developed in the amphibious *Ligia exotica* as compared with several completely terrestrial species (Takeda 1984). Holdich (1984) carried out a detailed scanning electron microscopic survey of possible cuticular receptors in a range of species of isopods, including amphibious forms as well as terrestrial species inhabiting mesic and arid habitats. All species possess tricorn-like setae, which are considered a probable terrestrial adaptation. Species inhabiting wet environments, including the amphibious forms, have open-ended (foraminate) tricorns while those from drier habitats have the setae enclosed within protective plaques. These setae are unique to oniscid isopods and may serve as sensillae that are involved with either or both hygroreception and mechanoreception.

Rittschof and Sutherland (1986) and Wellins et al. (1989) show that land hermit crabs (*Coenobita rugosis*) and the phylogenetically distant, semiterrestrial ghost crabs (*Ocypode quadrata*) both can detect and are attracted to a variety of volatile odor sources that indicate the presence of food. The sensillae responsible for this ability are located on the antennules in the hermit crab and apparently on the dactyls in the ghost crabs. Dactyl hairs of the ghost crabs are also involved in contact chemoreception of food-related substances (Trott and Robertson 1984). The semiterrestrial fiddler crab *Uca pugilator* uses contact chemoreception in its food gathering, but it responds primarily to a set of chemicals (carbohydrates) that are relatively unimportant as food cues for aquatic crustaceans. The latter respond primarily to amino acids (Robertson et al. 1981).

8.3.6 Reproduction

Water-land transitions have necessitated a variety of reproductive adaptations in the different crustacean groups. Powers and Bliss (1983) provide a general review of this topic, with emphasis on three particularly relevant aspects: the relationship of mating to molting; brood care in species that retain their larvae; and larval dispersal in species that do not retain their larvae. Specific reviews focusing on the three major groups are by Friend and Richardson (1986) for amphipods, Warburg (1987) for isopods, and Adiyodi (1988) and Wolcott (1988) for the land crabs. Here again the animals use a wide range of evolutionarily developed alternative scenarios. (This is particularly the case for the land crabs.)

Amphipods Mating in amphipods occurs when a mature female carrying ripe eggs molts. Among the beachfleas and sandhoppers copulation takes place between pairs during amplexus (the copulatory embrace); the duration of amplexus mating in these groups is substantially shorter than the hours to days of amplexus that normally occur in the intertidal hyalid amphipods that are considered the ancestral group of the talitrids. Observations of mating in landhoppers have never been reported in the literature. This void in data likely owes to the very short periods of amplexus in this group. Thus shortening of durations of amplexus is thought to be an important trend in the evolution of terrestrial adaptations in the amphipods. Male landhoppers are also almost always smaller than females of the same species, which is further support for this interpretation because amplexus in the more aquatic groups involves the males carrying the females about for the duration of the behavior.

Nothing is known about how male amphipods find females preparing to molt and lay eggs. Possibly one or more pheromones may be involved. If this is so, it seems likely that the pheromones used in air would differ from those used in water.

All amphipods carry their eggs in brood pouches. The embryos develop directly, hatching as post-larvae. The numbers of eggs per brood tends to decrease with increasing terrestriality. This is largely due to progressive increases in egg sizes, with resultant increases in sizes of young released. Greater openness of the structures of brood pouches is associated with larger egg sizes. This is interpreted as an adaptation for maintaining egg aeration in the absence of the respiratory current present in aquatic amphipods.

There is limited evidence for long-term sperm storage in the brood pouches of females of nonovigerous beachfleas and landhoppers. The sperm are presumed to be long lived. This capacity would help ensure continuing reproductive activity by females in times of low population densities, or, as is sometimes the case in landhopper populations, when sex ratios are strongly skewed toward females.

There seem to be no consistent trends relating to degree of terrestriality with respect to durations or numbers of breeding seasons or whether or not given species are semelparous (breeding once in a lifetime) or iteroparous (breeding more than once). Beachfleas and sandhoppers may be year-round breeders in the tropics, with several generations per year. They are seasonal breeders elsewhere, usually with two generations per year. Landhoppers may be either year-round or seasonal breeders, usually with one generation per year but sometimes breeding only in years with particularly favorable conditions.

Isopods The much greater evolutionary diversity of the isopods is manifested in many aspects of their reproductive biology. Amphibious isopods, unfortunately, have not been studied in detail. Mating in isopods does not seem to be timed to molting by the females. Amplexus mating does not appear to occur. Females deposit eggs into their own marsupia, where they develop and hatch into larvae. Some fraction of the eggs do not hatch, and—at least in *Ligia oceanica*—the larvae that do hatch feed on the eggs remaining. In several species of the family Ligiidae the number of eggs per brood varies from fewer than ten to more than sixty; durations of gravidity range from one to three months; and breeding seasons (in the temperate northern hemisphere) extend from late winter to early fall. Reproductive strategies have been partially investigated in only two amphibious species: one is semelparous, the other iteroparous.

Decapods The even greater evolutionary diversity of the decapods is also manifested in their reproductive biology. With respect to the *crayfishes* it is not clear whether or not there are any significant differences between amphibious and aquatic forms. The assumption appears to be that there are none.

There is great diversity among the *land crabs* in almost all aspects of reproduction. Few clear associations relating to degrees of terrestriality are, however, apparent. Mating generally occurs seasonally, with physiological phasing varying between species and, within species, between the sexes. Copulation may occur in water or in air. If copulation takes place in water, in many species it occurs between hard (intermolt) males and soft (postmolt) females, but many other cases are known of copulation occurring when both sexes are in intermolt (sometimes selective decalcification of the areas around the genital openings occurs during periods when mating is to take place). Copulation in air apparently occurs only between intermolt animals. Quite often courtship displays of considerable complexity precede mating. These can involve tactile, visual, and auditory signals. There is evidence that female sex pheromones are also involved in some cases (Dunham and Gilchrist 1988).

Amphibious crabs in general and a substantial proportion of terrestrial crabs release their eggs into bodies of water, most often the ocean. Most crabs showing this pattern of reproduction are considered to have evolved from marine ancestors. Varying degrees of direct development occur in terrestrial crabs that do not release eggs into water, some having no larval stages outside the egg. All but a few species of crabs showing this pattern are considered to have evolved from freshwater ancestors. Ovigerous females of the forms having larval development and dispersal in water often carry out extensive and presumably hazardous overland migrations to release their larvae. Larval metamorphosis in these species involves emergence onto land during or shortly after the megalops stage.

Land crabs generally are iteroparous. Especially in warmer climates they are likely to spawn several times during the breeding season. Reproduction in fully terrestrial species, particularly those specialized to live in trees, is likely to be triggered by seasonal patterns of rainfall (Dunham and Gilchrist 1988).

8.3.7 Behavior and Ecology

Each of the preceding sections on crustaceans has presented some significant information relating to the broad topics of behavior and ecology. We will not repeat any of that information here. Instead, we briefly summarize what is known about a few remaining topics not yet considered.

One of the most striking, and apparently nearly universal, properties of crustaceans is the rhythmicity in their activity patterns. Rhythms include circadian forms, plus those of tidal cycles, lunar cycles, semimonthly or monthly cycles, and seasonal or annual cycles. The shorter duration cycles, at least in decapods, are based on internal physiological pacemakers that determine approximate periodicity and which are entrained by environmental variables that fine-tune the phase relations. To date no unequivocal evidence has been obtained for endogenous controls on longer duration rhythms (Naylor 1988).

Amphibious and terrestrial crustaceans in each of the groups we are considering—amphipods, isopods, and decapods—demonstrate rhythmic behaviors at each of the time scales mentioned. The physiological mechanisms underlying these behaviors are almost certainly the same as those at work in aquatic crustaceans (indeed, amphibious crabs of the genus *Uca* have been the subjects of many of the most important studies elucidating these mechanisms). However, the ways in which these mechanisms are manifested are highly specific and adaptive parts of the ecological relationships underlying the survival and success of the water-land transitions in these groups. Manifestations include: tidally correlated movements up and down open sandy beaches by beachfleas and sandhoppers; circadian rhythms of feeding activity and the use of burrows and other refugia during stressful environmental conditions by amphibious isopods and many types of shore and land crabs; and the overland spawning migrations of many terrestrial land crabs, as mentioned in the previous section. Chelazzi and Vannini (1988) edited a recent symposium that summarizes a great deal of the relevant literature on these and related topics (e.g., spatial orientation mechanisms of beach amphipods and various intertidal decapods).

Substantial attention has also been devoted to the general issue of limiting factors in the environment that might restrict the distributions of crustaceans on the land. The two most widely cited factors, for all groups, are the availability of liquid water and inorganic ions— especially sodium, chloride, calcium, and (particularly among the isopods) copper. We believe that no hypotheses regarding environmental limiting factors have yet been confirmed. The evidence is at best circumstantial, largely hinging on the classic issues of the extent to which correlations imply causations and the extent to which other biotic factors, perhaps indirectly associated with the abiotic factors considered, may be the direct causes. These underlying, almost philosophical issues may never be resolved, but nearly all major reviews relevant to our subject matter discuss the problems of accurately determining causation.

8.4 Summary

Overall, it is apparent that both the land leeches and the crustaceans have been very successful in invading the land. It is also apparent that these successes are, to date, relatively limited in terms of phylogenetic diversity of the terrestrial groups (in comparison with their aquatic antecedents and contemporaries) and in terms of the range of environments that they occupy. The factors that have produced these limitations are not clear, though many hypotheses have been advanced over the years.

Strong circumstantial evidence is available for each major group— land leeches, amphipods, isopods, and decapods—concerning their probable phylogenetic aquatic ancestors and the most likely general environmental pathways followed in making the actual transitions. Direct invasions from marine habitats (including rocky coasts, sandy beaches, and estuarine habitats like mangrove forests and salt marshes) occurred, as did indirect invasions from freshwater habitats of many kinds.

The most probable patterns of biochemical, physiological, behavioral, and ecological mechanisms that were involved in the actual transitions are unknown and, within fairly broad limits, probably unknowable. The diversity of adaptive patterns (alternate scenarios) demonstrated by living forms is so great, and the extent of our knowledge of the relative frequencies of occurrence of these alternatives so limited, that we do not believe it is currently justifiable to select particular patterns as more likely to be ancestral than other patterns. At the same time it is clear that some features in each group are persistent preadaptations still present in extant members of the aquatic ancestral groups, while other features are new developments that presumably were perfected during the courses of the invasions. It is also clear that the transition scenarios now in use, probably for each major group, involve virtually all organ systems, as well as a wide variety of processes at the organ system level. Water-land transitions, like most other major evolutionary developments, have thus involved selective pressures and adaptational changes at most, if not all, levels of structural complexity—from the cellular to the organismic.

9

Functional Evidence from Living Vertebrates

MALCOLM S. GORDON

In this final chapter exploring what can be learned of the water-land transition from functional evidence available in living metazoans, we focus on the vertebrate members of the phylum Chordata. Among the extant vertebrates, the amphibious osteichthyan fishes and the amphibians are the relevant groups. We omit the reptiles. The reptilian final stages of the water-land transition were very complex and many aspects remain uncertain. We therefore focus on the somewhat clearer early parts of the story.

The many entirely aquatic but air-breathing fishes are also omitted. The primary reason is that these animals show no adaptations for life out of water other than their ability to breathe air. Their respiratory capacities probably evolved in response to selective pressures only partly related to water-land transitions. The main ecological result is that these particular lineages have been able to remain aquatic in the face of hypoxic and anoxic circumstances in a variety of (mostly freshwater) habitats.

9.1 Scope of Study

The starting points for this discussion include the chapter on vertebrates in Little (1983) and the relevant sections in Little (1990). Because the subject matter of this chapter has been and remains one of the most popular topics in vertebrate biology, there are several additional reviews that are both recent and general. These include: Bond (1979), Duellman and Trueb (1986), and Pough et al. (1989).

The recent, nearly encyclopedic *Environmental Physiology of the Amphibians* (Feder and Burggren 1992) provides a series of overviews of the literature on the living Lissamphibia with respect to many subjects also considered here. It includes a bibliography of about 4,000 references that is almost comprehensive within its areas of coverage through 1989. The theoretical contexts and concerns upon which most of the chapters in Feder and Burggren (1992) are based are, however, markedly different from those of this chapter. Notably, we emphasize matters relating to water-land transitions and omit adaptations to other environmental situations. In addition, substantial numbers of significant papers have appeared since the time that Feder and Burggren and their coauthors completed their manuscript, and we refer to those papers in our discussion here.

Of all the living groups of animals included in this book the vertebrates are the only group in which it is possible to specifically and unequivocally discuss each of the three important categories of organisms relevant to water-land transitions: the fossils; living forms considered to be phylogenetically related to fossil forms thought to have been involved in the actual transitions; and living forms currently occupying habitats and having life histories that might reasonably be considered similar to the places and ways in which the actual transitions might have occurred. Vertebrates are the best group in which to study the water-land transition because the morphological differences between fishes and amphibians are usually so large that fossil members of the groups on both sides of the transitions are readily recognizable. An important consequence is that for the vertebrates clearer statements can be made than are possible for the other groups with respect to which characteristics of extant terrestrial organisms probably were derived from preadaptations possessed by putative aquatic ancestors and which were developed as part of the water-land transition itself.

Because of this fortuitous characteristic of vertebrates this chapter can, in many respects, be more selective and less comprehensive than the previous two chapters. Quite simply, we know where to look. In addition, the total biodiversity of amphibious fishes and true amphibians is greater than that of the relevant segments of the other groups, and much larger fractions of that biodiversity have been studied in various ways. The literature is therefore even larger and less susceptible to clear, compressed summarization.

The higher classification of the vertebrates continues to be the subject of ongoing research and disputation. In this book we adopt a more or less middle-of-the-road position that we consider consonant with the majority of the evidence. This is broadly reflected in the listing of groups with terrestrial representatives given in table 4.1. The names of groups we will use here are occasionally different from those of table 4.1, as we derive the specifics from Nelson (1984) for the fishes and from Duellman and Trueb (1986) for the amphibians. We now list the groups to be discussed, with equivalences listed in table 4.1 given in parentheses.

All fishes relevant to this discussion belong to the class Osteichthyes and to four subclasses: Dipneusti, Crossopterygii, Brachyopterygii (all three in subclass Sarcopterygia) and Actinopterygii (subclass Actinopterygia). For reasons discussed in chapters 5 and 6, we omit the lungfishes (subclass Dipneusti, order Dipnoi). The living coelacanth, *Latimeria*, is a part of the story; it belongs to the crossopterygian order Coelacanthiformes.

With one exception, all living amphibious fishes, both marine and freshwater, are teleost actinopterygians. The single exception is the brachyopterygian polypteriform reedfish, *Erpetoichthys calabaricus* (Sacca and Burggren 1982). At least five orders of actinopterygians contain amphibious forms: Anguilliformes, Siluriformes, Gobiesociformes, Synbranchiformes, and Perciformes. Within these orders a total of about ten families, each including a substantial number of genera, contain amphibious species. Three of those orders include only one family; the Siluriformes include at least three families, the Perciformes at least four. The total number of amphibious species is unknown, but it is probably between 100 and 150.

The classification of the fossil Amphibia is unresolved (see chapters 5 and 6; Duellman and Trueb 1986:494). All living Amphibia belong to the subclass Lissamphibia, containing three orders: Caudata, Gymnophiona, and Anura. The living Caudata include nine families, 62 genera, and about 350 species. The living Gymnophiona include six families, 34 genera, and about 160 species. The living Anura include 21 families, 301 genera and about 3,450 species. The higher level taxonomies of each order are not well established. Substantial evidence for polyphyly exists. Life histories are very variable, especially in the Anura. Some forms are completely aquatic; others are fully terrestrial, fossorial, or arboreal.

The ready identifiability of *Latimeria*, of amphibious fishes, and of amphibians makes it possible for us to discuss separately those properties relevant to the water-land transition that occur in: (1) the sole living representative (*Latimeria*) of the group closest to the evolutionarily ancestral group; (2) an array of living species that are clearly on the water side of the transition (amphibious fishes); and (3) another array of living species that are on the land side (amphibians). This sequence will be followed in each of the subsections.

The extents to which *Latimeria* and the amphibious fishes are representative of the larger groups—throughout their whole evolutionary history—to which they belong cannot be determined. There is only one species of coelacanth alive. The number of species of amphibious fishes is tiny compared with the overall diversity of osteichthyan fishes, which probably number more than 25,000 species. And as noted earlier, the amphibious fishes are themselves phylogenetically diverse. This makes it probable that amphibious traits have evolved separately at least several times. Relatively few types of amphibious fishes have been studied in detail. Thus the present picture of adaptive scenarios used by the amphibious fishes is incomplete and potentially misleading. This set of scenarios does provide an envelope of possibilities with respect to how the actual transitions might have occurred at different times and places. There is, however, no scientific basis for inferring that any one of these scenarios is more or less evolutionarily significant than are the others.

The summaries presented in this chapter are intended to provide relatively broad and general overviews of the diversity of possible adaptive scenarios as shown by both the amphibious fishes and the true amphibians. These summaries are not intended to be detailed critical reviews of the entire literature on each of the major subjects addressed. Our hope is that selective considerations of better known cases will more effectively illustrate the principles and themes than will welters of detail.

The best recent source of information concerning *Latimeria* is Musick et al. (1991). Gibson (1982, 1986) includes some information on amphibious fishes in his two reviews of the biology of intertidal fishes. Sayer and Davenport (1991) consider why amphibious fishes emerge from the water. Duellman and Trueb (1986) and Feder and Burggren (1992) exhaustively survey the amphibian literature.

9.2 Mechanical Support, Locomotion, and Functional Morphology

Large parts of the literature of vertebrate morphology, both descriptive and functional, are devoted to the myriad consequences for basic body plans—and especially for musculoskeletal and nervous systems—that can be attributed to the loss of mechanical support, and the consequent exposure to the direct forces of gravity, that was a fundamental challenge of the water-land transition. Recent reviews are by Hildebrand et al. (1985), Radinsky (1987), Bramble et al. (1989), and Gans and deGueldre (1992).

The major features of the anatomical arrangements and the structures of tetrapod limbs are probably examples of preadaptations first developed in completely aquatic fishes. It is also probable that the basic neural and neuromuscular circuitry for producing the movements of these limbs, plus substantial portions of the motor and sensory programming involved in the control and coordination of those movements, also evolved in aquatic fishes. There are two lines of evidence supporting this position. First, submarine observations of living *Latimeria* show that the paired fins of slowly swimming fishes move synchronously and alternately in a tetrapod mode of locomotion (Fricke et al. 1987; Fricke and Hissmann 1992). Second, Edwards (1989) presents evidence that several types of living, benthic bony fishes that are heavier than sea water and that move along the bottom by walking do so in modes that are strongly convergent with the walking modes of terrestrial sprawling tetrapods (e.g., salamanders and lizards). These convergences make it at least plausible, if not probable, that the physical and mechanical constraints associated with overland pedestrian locomotion are so strong that the vertebrate solution to these problems first evolved underwater (see also Frolich and Biewener 1992).

In this context it is interesting to note that *Latimeria*, while apparently a benthic fish, has a body containing large quantities of lipids that make it neutrally buoyant (Nevenzel et al. 1966). Thus it moves about by swimming, not walking. There is, of course, no way of determining whether or not the same was true of the ancestral crossopterygians.

The living amphibious fishes are almost all effective in terrestrial locomotion. The different groups use a wide range of locomotory modes—all different from the tetrapod mode. Most forms use several different modes as need dictates. These include:

- sinuous, straight-ahead snakelike progression over smooth surfaces
- tripodlike walking, using the tail for balance and the pectoral fins for propulsion
- sidewinding, similar in appearance to that done by some sanddwelling snakes
- jumping, usually based on springing from a position with the tail strongly curled.

Endurance and speed vary more or less inversely in these different modes, with sinuous, straight-ahead progression being the slowest but most sustainable and jumping being the fastest but most rapidly fatiguing. Maneuverability is substantial in all modes (Brillet 1986;

Brown et al. 1991). It is apparent from this present diversity that the tetrapod mode of locomotion was not necessarily the only evolutionary option that might have been available for use by vertebrates leaving aquatic environments. The selective tradeoffs that resulted in the adoption of only the tetrapod option are unclear.

The aquatic larvae of living amphibians that have metamorphoses as important parts of their life histories are morphologically diverse. This diversity has a variety of implications for swimming modes and for both the energetics of swimming and the maneuverability of the different morphological forms. There also are a range of interactions between different forms and the speeds and sequences of morphogenetic processes during metamorphosis. Wassersug (1989) reviewed many aspects of these matters. Two important conclusions were that: (1) metamorphoses in anurans are strongly constrained in several respects and must be as brief as possible; and (2) the mobility implications of metamorphosis differ from group to group in ways that may reflect the phylogenetic histories of each. Dudley et al. (1991) provided additional support for the first conclusion. They showed that drag forces on swimming anurans rise abruptly and substantially at metamorphic climax (when all four legs are exposed, but the tail is still full). This increase in drag results in decreased endurance while swimming and increased vulnerability to predation.

Adult amphibians moving about on land also demonstrate substantial variability in their locomotory modes and patterns. As with the amphibious fishes, individual species may use any of several modes depending on circumstances. Serpentine wriggling, walking, running, and hopping are four of the major alternatives. The comparative energetics of different modes of locomotion are unstudied in the amphibious fishes, but such work has begun on amphibians in recent years (Pough et al. 1992; Gatten et al. 1992).

An important concern in some of this work has been to evaluate the relative contributions to overall metabolic demands in subaerial activity of cutaneous as compared with buccal and lung respiration. Four species of salamanders and one species of toad have been studied (Full et al. 1988; Walton and Anderson 1988). Factorial aerobic metabolic scopes (the ratios of maximal active to standard metabolic rates) were found to be only 1.6–3.0 in the lungless salamanders, but they increased to 3.5–7.0 in the lunged salamanders and reached 9 in the toad. These scopes are relatively low for terrestrial vertebrates generally. It was also discovered that the salamanders had lower aerobic costs of transport (60–80 percent lower) than did lizards of comparable body masses. Cost of transport in hopping toads is not significantly lower than costs of running in other animals of similar body sizes. Endurance times of lungless salamanders proved to be much shorter than those of lunged salamanders at given speeds, and maximal speeds were also substantially lower in lungless forms. Maximum endurance times at moderate speeds were twenty minutes or less for lungless salamanders, but one to two hours for lunged forms. Toads could hop for more than an hour at moderate speeds. Major conclusions were: (1) complete reliance on cutaneous respiration seriously limits the upper levels of performance in lungless as compared with lunged salamanders; and (2) hopping in toads, as in small mammals, is most adaptive during short bursts of high activity and is not economical in sustained locomotion.

Recent experimental approaches to specific aspects of the functional morphology of the tiger salamander (*Ambystoma tigrinum*) before and after metamorphosis shed light on what may have been some additional preadaptations for terrestrial life that first evolved in aquatic environments. Lauder and associates (Shaffer and Lauder 1988; Lauder 1990; Lauder and Reilly 1990; Reilly and Lauder 1990; Ashley et al. 1991) studied changes across metamorphosis that occur in tiger salamanders with respect to both pharyngeal (jaw and throat) and hindlimb musculatures. The interest in the former is that mechanisms used to transport prey from the jaws to the throat change from primarily hydraulic transport to primarily tongue-based transport between aquatic and terrestrial vertebrates. The hindlimb musculature was thought likely to be affected by the shift in gravitational loads that accompanies emergence onto the land. Important results were: (1) The kinematic profile of tongue-based prey transport in salamanders on land is very similar to that used by fishes and salamanders in water during aquatic prey transport. (2) Muscle activity patterns during aquatic feeding are constant across metamorphosis, but tongue-based terrestrial feeding involves a suite of novel muscle activity patterns. (3) Terrestrial feeding also involves a number of new morphological features acquired during metamorphosis, plus an increase in the masses of muscles important in tongue projection (see also Miller and Larsen 1990). (4) In contrast, hindlimb muscle masses scaled appropriately for changing body size do not change during metamorphosis. The growth trajectories of the hindlimb muscles established in the larva are extended with little or no change into early terrestrial life.

Combined, these brief observations about mechanical support and locomotion do not cover the totality of the differences that characterize both the descriptive and functional morphologies of aquatic as compared with terrestrial vertebrates. However, they do demonstrate the far-reaching nature of the changes that occurred and the extent to which important aspects of these morphologies may have developed well before the actual water-land transitions occurred. The very small numbers of living species used in the work that produced these apparent insights also should be viewed as cautionary. The degrees of generality of the results are not yet established. Thus we think it prudent not to extrapolate too far from these findings to sweeping speculations concerning their possible implications for the lives of the earliest tetrapods.

9.3 Respiration, Circulation, and Metabolism

No area relating to vertebrate water-land transitions has received, and continues to receive, as much research interest as does respiration and its closely related structures and processes. The transitions from gill to lung respiration were themselves highly complex, with additional complications owing to contributions from buccopharyngeal and cutaneous respiration in both fishes and amphibians. Respiratory structures, mechanisms, and controls have all changed dramatically. The processes of respiratory gas transport and of acid-base regulation in blood and tissue fluids have also all been substantially modified in the course of land colonization. The circulatory systems associated with the several organs of external respiration demonstrate substantial variability on both sides of the transition, as well as large-scale morphological adjustments across the transition. There have been many metabolic consequences resulting from all these changes.

Important general references on matters of respiration, circulation, and metabolism were already listed in chapter 8. The major structural features of fish and amphibian respiratory and circulatory systems are summarized by Bond (1979:41–49) and Duellman and Trueb (1986:397–405). (*Latimeria* is not a relevant organism in the context of this section. Since it is completely aquatic we omit it from further consideration.) Knowledge of amphibious fishes is still limited in terms of systematic coverage of the diversity of forms. Most work has been done with intertidal fishes (Bridges 1988). That review is supplemented by Hyde and Perry (1987), Martin (1991), Brown et al. (1992), and Yoshiyama and Cech (1993).

9.3.1 Amphibious Fishes

Many amphibious intertidal fishes spend almost their entire lives out of water—breathing air cutaneously, buccopharyngeally, and across their gills. These fishes, however, will not drown if kept for extended periods in water without access to air (Bridges 1988). The relatively brief contacts with liquid water that are from time to time made by these animals serve not only to keep their skins and gills moist but may also allow them to replenish the small volumes of water many of them carry about with themselves, in their gill chambers, while they are on the land. These water volumes may play important roles in aerial respiration as well as in blood acid-base balance and waste nitrogen excretion (Martin and Lighton 1989; Brown et al. 1991).

Gill morphologies and morphometries are variable between species, with some correlations between degrees of terrestriality and gill structures. The more terrestrial species may have shorter gill primary lamellae, and shorter and less closely spaced secondary lamellae. Considerable interlamellar fusion occurs in some species. These features may combine to produce relatively smaller surface areas of the gills than characterize fully aquatic fishes of similar sizes and activity patterns. The gills of the more terrestrial forms also often have better developed cartilaginous gill rods (which help support the primary lamellae and keep them separated from one another) than do more aquatic species. A number of species have specific regions of increased vascularization in their skins, which are probably accessory respiratory structures (Brown et al. 1992).

Ventilatory movements of the gill opercula typically occur when amphibious fishes are in water. Such movements are usually infrequent and sporadic in most species when out of water. Some of the most terrestrial forms (e.g., the most terrestrial species of gobiid mudskippers) spread their opercula after taking a mouthful of water, then make no further respiratory movements until their next visit to water. Other species (e.g., several blennies) periodically flutter their opercula for short periods; this behavior may facilitate carbon dioxide release (Brillet 1986; Martin and Lighton 1989).

Routine rates of oxygen uptake by amphibious fishes are usually about the same, whether in or out of water, but some species deviate substantially from this norm. Some show notably higher rates of uptake in air than in water, others much lower rates (Bridges 1988; Brown et al. 1992).

There is little indication that most species carrying out routine activities out of water are significantly stressed metabolically. The conditions under which these animals show strong signs of either or both respiratory and metabolic acidosis are usually those involving higher levels of locomotory activity (eels of the genus *Anguilla* may be important exceptions to this; Hyde, Moon, and Perry 1987; Hyde and Perry 1987). Major features of the "diving syndrome"—but in the reverse direction (from water to land)—that is often associated with breath-hold diving (e.g., lowered aerobic metabolic and heart rates, plus substantial tissue accumulation of lactic acid during dives) occur to varying extents and in different combinations in different species of amphibious fishes (Bridges 1988; Martin 1991).

A small amount of interesting information is available about a few aspects of respiration and metabolism in one species of freshwater polypterid fish that is amphibious. The reedfish (*Erpetoichthys calabaricus*), under laboratory conditions, makes several spontaneous excursions onto land each day— each excursion lasting for one to three minutes. However, it can tolerate several hours in air. Oxygen exchanges occur by way of the gills, the skin, and the pair of large lungs. Oxygen uptake rates in air are higher than uptake rates in water. These increases are due primarily to increased lung uptake, but also to limited increases in cutaneous uptake (Sacca and Burggren 1982). Lung ventilation in this fish, as well as in the completely aquatic *Polypterus*, occurs by recoil aspiration, based on the deformation and recoil of heavily bony-scaled integuments. These are the only lower vertebrates that are known to use aspiration breathing. An interesting structural and functional parallel with the bony-scaled integuments may be found in some Paleozoic (Carboniferous) amphibians, which had ventral, bony scales (Brainerd et al. 1989). It is therefore possible that recoil aspiration ventilation may have been important to some early tetrapods.

A summary observation concerning amphibious fishes in the present context is that, in at least the most important respects, all of the physiological capacities required for survival and fairly extensive functioning on land exist and are quite well developed in a variety of forms. It is likely that the different sets of adaptations used by the different groups evolved independently of one another.

9.3.2 Amphibians

Amphibian respiration, circulation, and metabolism are the basis for a very large literature. All important aspects have received significant attention in recent times. There is substantial diversity in structures, mechanisms, and controls within and between the major groups. Many different adaptive scenarios for the water-land transition are viable hypotheses, and once again there is no sound basis for choosing among them. Recent articles not cited in Boutilier et al. 1992 or Shoemaker et al. 1992 include: Wood and Glass 1991 (respiration in general and acid-base balance); Liem 1988, Vitalis and Shelton 1990, and Feder and Booth 1992 (lung and cutaneous respiration); Milsom 1991 (respiratory control); Burggren and Pinder 1991 and Hillman 1991 (circulation); Flanigan et al. 1991 and Gleeson 1991 (metabolism). Limited and fragmentary information is available relating to caecilians. The most recent substantial publications are Toews and MacIntyre (1977, 1978).

Patterns of lung breathing out of water in most amphibians that have been studied are different from the sinusoidal breathing characteristic of birds and mammals. Amphibian lung breathing is instead episodic, with irregular ventilatory periods interrupted by variable intervals of apnea. This difference in pattern has many significant implications with respect to: sensitivities of animals to hypoxia or hypercapnia; mechanisms of respiratory control; and cardiovascular relations to respiration (Milsom 1991).

Amphibians possess at least one pattern of external respiration that does not exist among the amphibious fishes: the solely cutaneous and buccopharyngeal respiration of the lungless salamanders. All other amphibians respire across their skin and lungs (adult urodeles and anurans) or use gills and skin (most larvae) or a combination of gills, skin, and lungs (neotenic larvae). Excepting the lungless salamanders, all forms have complex and effective mechanisms for adjusting both cardiac outputs and patterns of blood flow to changing respiratory needs (Burggren and Pinder 1991; Hillman 1991).

The physical implications and the metabolic limitations that result from substantial or complete reliance on cutaneous respiration have only recently begun to be more fully understood. Cutaneous respiratory exchanges by amphibians in water are greatly affected by the presence of unstirred diffusional boundary layers adjacent to their skins. Active movements by the animals that cause mixing, or the presence of water flows in the environment, significantly enhance rates of exchange for both oxygen and carbon dioxide. The presence and properties of these boundary layers have many important consequences for both the behavior and ecology of amphibians in water. Amphibians in air are not substantially affected by this problem (Vitalis and Shelton 1990; Feder and Booth 1992).

Overall rates of oxygen uptake in amphibians are often limited by the capacity of the cardiovascular system to convectively transport oxygen. Oxygen uptake is not limited by diffusion in either the organs of external respiration or the body tissues. The principal cardiovascular bases for variations in rates of oxygen uptake due to physical activity are changes in arteriovenous oxygen content differences and changes in heart rates. Stroke volumes, in contrast, are little changed during increased physical activity. Dehydrational water losses from amphibians on land cause cardiac hypovolemic stresses. The ability to compensate for these stresses is an important adaptation in more dehydration-tolerant forms (Hillman 1991).

The major respiration-related features of the skin in those amphibians that have been studied are surprisingly constant across metamorphosis. This is despite often dramatic changes in both gross appearance and function (Shoemaker et al. 1992).

Finally, laboratory evidence from a few species of adult anurans indicates that exposures to environmental hypoxia induce the animals to seek environments with lower ambient temperatures. Hypoxic stresses in most amphibians result in increased rates of anaerobic metabolism. In addition, the species studied appear to further reduce their oxygen demand by behaviorally lowering their body temperatures (Hutchison and Dupré 1992).

9.4 Thermal, Water, and Solute Relations

Direct effects of temperature on amphibious fishes and amphibians are similar and comparable to those found in ectothermous animals generally. There are, however, several ways in which the change in environments has dramatically affected many of the consequences of temperature variations. All of these consequences include important effects on water and solute relations.

For example, the most significant and pervasive change associated with the transition is undoubtedly that of exposure to unrelenting water losses in most terrestrial environments. The magnitudes and rates of these losses are usually large, and they are strongly affected by temperature variations and variations in both convective and radiative exchanges with the surroundings. Many different adaptive mechanisms and responses have evolved to permit amphibious animals to cope with these stresses. (Aquatic and burrowing forms, also species living in very moist environments, are often exceptions to this situation.) In turn, the often greater thermal complexity of terrestrial environments makes it possible for amphibious vertebrates to choose microhabitats. Behavioral thermoregulation and increased thermal tolerance have thus both been selected for. Thermal acclimation, however, appears relatively less important, at least in the amphibians. Seasonal variations in temperature, particularly in temperate zone and subarctic habitats, have also selected for adaptations. These include both seasonal torpidity, most often also associated with burrowing, and direct tolerance of physical freezing.

Sayer and Davenport (1991) consider thermal, water, and solute relations with respect to amphibious fishes. Several other papers also provide such information (Horn and Riegle 1981; Gordon et al. 1985; Hyde, Moon, and Perry 1987; Hyde and Perry 1987; Brown et al. 1991). The literature specific to amphibians is reviewed in Rome et al. 1992, Hutchison and Dupré 1992, Boutilier et al. 1992, and Shoemaker et al. 1992. Other particularly relevant articles are (for temperature) Storey and Storey 1988, Wollmuth and Crawshaw 1988, and Wood and Glass (1991); (for water and solutes) Hillman 1988, Hanke 1990, Stiffler et al. 1990, and Brekke et al. 1991.

9.4.1 Water Losses

All amphibious fishes that have been studied usually remain close enough to the water's edge and are sufficiently active and mobile that it is improbable that evaporative water losses are a significant hazard to survival under natural conditions (Brown et al. 1991). However, those species that have been specifically studied can tolerate dehydration to extents that are comparable with tolerance limits for many semiterrestrial anurans. Rates of evaporative water loss under comparable conditions of temperature, vapor pressure deficit, wind velocity, and solar insolation are usually somewhat lower than those shown by anurans. Dehydrational tolerance limits usually vary inversely with rates of water loss—higher tolerance being associated with slower desiccation. This relationship implies that causes of death from dehydration are likely to be factors other than simply changes in levels of body water content. What those factors may be is unknown (Gordon et al. 1978).

Different groups of amphibians show great variability in tolerance limits for desiccation. The caecilians appear to be unstudied in this respect, but since they are all either aquatic or burrowers desiccation may be presumed a nonproblem (Stiffler et al. 1990). Aquatic and burrowing caudates and anurans also generally avoid these problems, but those burrowing forms that live in xeric habitats and survive long periods of adverse surface conditions through torpidity underground necessarily dehydrate slowly as the soils surrounding their refugia dry out. Some of these animals can tolerate losses of 50 percent or more of their body weight under these conditions. Some species produce dry, parchmentlike cocoons that encase their bodies, leaving only nostrils and/or mouth exposed. These cocoons derive from the outer layers of the skin. They serve to restrain rates of water loss to levels well below those shown by animals without cocoons subjected to similar conditions (Shoemaker et al. 1992).

Most other aquatic, amphibious, or terrestrial amphibians have skins highly permeable to evaporative water loss. The skins of various species that have been studied evaporate water at rates comparable to those of free water surfaces of similar area under the same environmental conditions. There are, however, important differences among species in terms of rates of osmotic rehydration once free water becomes available again in the environment after a period of desiccation (drinking of liquid water appears rare in amphibians). Aquatic and amphibious species rehydrate much more slowly than do terrestrial or arboreal species. The latter groups are also likely to have specific areas of skin (usually pelvic patches) that are specialized for rapid water uptake (Shoemaker et al. 1992). Considerable attention has been paid to the mechanisms of rehydration by investigators interested in comparative endocrinology. The primary hormone regulating cutaneous water uptake is usually arginine vasotocin, but recent evidence indicates that at least some desert toads may also use angiotensin II (Brekke et al. 1991).

Major exceptions to the physiological responses just described are the so-called waterproof frogs belonging to at least three families of tree frogs that live in xeric areas in Africa, South America, and Australia. To varying degrees the dorsal surfaces of these animals show greatly reduced rates of evaporative water loss. Measured resistances to cutaneous water vapor losses in the most highly adapted species are comparable with resistances measured in several species of lizards living in drier environments. Different mechanisms produce these low permeabilities in the different groups. Specializations in skin structure seem important in some, while others produce waxy secretions from unusual skin glands. The latter have behaviors that actively spread the gland secretions over the exposed body surfaces. In all groups of waterproof frogs, organisms exposed to full midday sun and wind will often spend extended periods of time resting in postures serving to cover their underbody surfaces—which show higher permeabilities to water vapor (Shoemaker et al. 1989; Shoemaker et al. 1992).

The diversity in adaptations just described makes it at least possible that there may have been similar diversity in adaptations for dealing with dehydration among the ancestral amphibious fishes and Amphibia. The first stages of the vertebrate water-land transition necessarily occurred across some shorelines somewhere, but those shorelines need not have been restricted to marshy or even mesic terrestrial habitats.

9.4.2 Thermal Relations

Behavioral thermoregulation appears almost absent from the amphibious fishes—at least the few species that have been studied in some detail. This absence may be associated with a general lack of selection for such capacities. Almost all species spend most of their time close to water and most make frequent visits to liquid water. Preferred body temperatures determined under laboratory conditions can, however, be substantially different from, and vary over narrower ranges than, body temperatures measured in the field. Body temperatures measured in the field are often close to acute upper lethal temperatures. In general, factors other than temperature (e.g., food availability, or cover from potential predatory attack) seem more important in determining choices of microhabitat made by amphibious fishes under field conditions. However, when high ambient temperatures combine with clear skies during midday, shade-seeking or burrow-entering behaviors for thermal reasons may well occur in at least some species (Brown et al. 1991; Sayer and Davenport 1991).

Amphibian thermal relations have been extensively studied, and they are often complex. Thermal properties of aquatic amphibians are usually much like those of fishes living in similar habitats. Multiple interactions between environmental conditions and organismic response occur in amphibious, terrestrial, and arboreal forms. Examples include interactions among ambient temperatures, evaporative cooling, convective air movements, conductive exchanges with the substrate, and radiative exchanges with the surroundings and the sky. These interactions can be modified by a range of other factors, including nutritional state, acclimation temperature, developmental stage, reproductive state, time of day, and availability of microhabitat choices. Observed body temperatures in the field vary widely diurnally, seasonally, and geographically. Preferred body temperatures measured in the laboratory may or may not be similar to field values (Rome et al. 1992; Hutchison and Dupré 1992).

An interesting interaction that may have survival value for amphibians in thermally complex environments (e.g., thermally stratified ponds) is a shift to lower preferred body temperatures when the organism is exposed to significant environmental hypoxia. This shift (recorded in the few species thus far studied) may result in lowering aerobic metabolic needs, thus improving chances for surviving periods of low oxygen availability (Wollmuth and Crawshaw 1988; Rome et al. 1992; Hutchison and Dupré 1992).

The waterproof frogs mentioned earlier show substantial capacities for physiological regulation of their body temperatures when ambient temperatures approach or exceed their upper thermal limits. Body temperatures in these animals stop tracking ambient temperatures at these higher ranges. The basis for this regulatory ability is increased evaporative cooling owing to increased rates of mucus secretion by many skin glands. This mechanism is thus analogous to sweating in mammals (Shoemaker et al. 1989; Shoemaker et al. 1992).

9.4.3 Seasonal Adaptations

We discuss here only two of the many adaptations of amphibious vertebrates that vary seasonally: torpidity and freezing resistance. As far as is known, neither adaptation occurs in amphibious fishes. It is possible, however, that some higher latitude intertidal fishes (e.g., cottid sculpins, some blennies, and stichaeids) that routinely spend time out of water, most often in the cover of mats of seaweed, may be exposed to and tolerate subfreezing temperatures in winter. If this does happen, some of these fishes may also survive partial or complete freezing of their body water.

Seasonal torpidity occurs widely among amphibians. Torpidity is distinguished by persistent inactivity and lowered metabolic rates. It apparently occurred in some of the oldest known fossil forms (Carboniferous). In living species torpidity is usually associated with burrowing—which may be into mud under or adjacent to a stream or pond, or into the soil of a forest, or into sandy soils in xeric habitats or deserts. It may be triggered by low winter temperatures, high summer temperatures, or a range of combinations of temperatures and dryness. Durations of torpidity may be just a few weeks or many months. Whatever the cause or the duration, there are several uniformities in what the animals entering torpidity actually do: they become inert and relatively unresponsive to external stimuli; their metabolic rates decline substantially; and they cease voiding their urine, using their bladder water as an important basis for internal water recycling and waste nitrogen storage (Shoemaker et al. 1992).

A more detailed and complete picture of patterns and mechanisms of regulation of body water and solutes during winter hibernation derives from recent work on European toads (*Bufo bufo*). Temperate zone anurans hibernating at low temperatures in moist surroundings characteristically slowly become edematous. In their hibernacula, they begin to take up from their surroundings, and to retain, substantial amounts of both water and NaCl, with resulting large increases in the volumes of extracellular fluid in their bodies. Much of this fluid is held in distended lymph sacs.

The physiological processes inducing the edematous condition include reduced diuresis and relatively high cutaneous water uptake. The increased body salt content results from increased net cutaneous salt uptake, which begins a few days after the start of water uptake. The rates of both water and salt uptake increase over a period of about ten days, then remain high for weeks. A new steady state develops after several months. When higher temperatures return in spring, these water and salt loads are rapidly eliminated via the kidneys. It appears that the patterns of change are brought about by resettings of set points for body extracellular volume and ionic concentrations (therefore also body fluid osmotic pressures). The changes are not due to some kind of defect in osmoregulatory mechanisms that may be induced by low temperature (Hanke 1990; Pinder et al. 1992).

Temperate zone and subarctic species of anurans that live in places where winter temperatures in their hibernacula fall below the equilibrium freezing temperatures of their extracellular body fluids first supercool to small extents, then physically freeze. If exposed to air while frozen they will freeze-dry within days and die. Under more normal conditions (buried under damp leaf litter or mats of moss, plus some overburden of snow) they readily survive extended periods at the lowest microhabitat temperatures that are likely to occur in their ranges (no lower than $-7°C$).

At least five species of North American frogs have been shown to tolerate extracellular freezing under these conditions (based on freezing point depressions of their body fluids of near 1°C, at $-7°C$ over 80 percent of body water should be frozen). As the freezing ensues, the livers of the frogs synthesize from glycogen significant amounts of either glucose or glycerol. These are released into the body fluids, where they act as partial cryoprotectants. Body fluids pool and freeze in distended large blood vessels above the heart, as well as in tissue spaces elsewhere. The tissues survive total ischemia while frozen by using low rates of anaerobic metabolism. Lactate and alanine accumulate as end products.

When thawed, first signs of life are the return of the heartbeat, then gulping movements. Experiments in the laboratory show that with gradual thawing over many hours at low temperatures like those probable under natural conditions, muscle tension returns to the limbs only near completion of the process. Voluntary motor movements then become possible. Freezing tolerance persists for some time during early spring breeding, but it declines sharply later in spring after the animals resume feeding (Storey and Storey 1988; Pinder et al. 1992; Costanzo et al. 1992; Storey and Storey 1992).

The diversity of adaptations related to temperature, water, and solutes that exists in the living amphibious vertebrates—only some of which were presented here—once again make it possible, if not probable, that comparable diversity occurred among the ancestral amphibious fishes and Amphibia. There is thus no reason to restrict the geographic locus of vertebrate transition scenarios to paleoenvironments of the humid tropics. Nor is it wise to restrict the geographic locus to freshwater shorelines. The existence of at least several species of moderately to strongly euryhaline amphibious fishes and amphibians means that brackish water environments, such as mangrove forests and salt marshes, might also have served as routes for water-land transitions. As we mentioned in chapter 5, no fossil vertebrates have been found that appear to be animals directly involved in the transitions; the actual environment or environments in which the events took place must thus remain speculative.

9.5 Nitrogenous Metabolism and Excretion

Amphibious vertebrates in water—whether marine, brackish, or fresh—usually eliminate substantial fractions of their nitrogen wastes as either ammonia or ammonium ions that diffuse or exchange for other ions across their gills or skins. Many amphibious fishes, like substantial numbers of completely aquatic bony fishes, also produce some urea, most of which is filtered out by the kidneys and eliminated with the urine (Walsh and Henry 1991).

Out of water, or at least in laboratory situations in which the organisms are prohibited access to liquid water, amphibious fishes usually cease voiding urine and shift the balance between metabolic ammonia and urea production fairly strongly toward urea. This appears to reduce the possibility of ammonia toxicity under these conditions. When returned to water these animals quickly eliminate the extra urea loads, probably primarily via the urine (Gordon et al. 1978; Iwata 1988; Ip et al. 1990).

Completely aquatic amphibians are usually similar to aquatic fishes. However, when faced with adverse environmental conditions involving severe water restrictions, such as seasonal drying of the ponds in which they live (which may then induce burrowing and torpidity) or increases in ambient osmotic concentrations as ponds dry out, otherwise aquatic amphibians can shift toward ureotelism and accumulate significant amounts of urea in their body fluids (Boutilier et al. 1992).

Amphibious, terrestrial, and arboreal amphibians shift to ureotelism when out of water. However, the waterproof frogs have evolved the capacity to conserve larger amounts of urinary water by excreting most of their waste nitrogen, and excess inorganic salts, in the form of uric acid and urates (Shoemaker et al. 1992).

Waste nitrogen metabolism and excretion by anuran tadpoles, or the embryos of anurans with direct development, have been studied in only a few species. Aquatic tadpoles are ammonotelic and lack urea cycle enzymes in their livers. These enzymes only appear during the later stages of metamorphosis, producing the ureotelic capacities just described. Nonaquatic tadpoles and direct developing embryos apparently synthesize the urea cycle enzymes before metamorphosis. Tadpoles of at least some of the South American waterproof frogs also synthesize uric acid and urate salts (Boutilier et al. 1992; Shoemaker et al. 1992).

Some species of both caudates and anurans have been found capable of accumulating urea in their body fluids (blood, lymph, urine, and presumably most other fluid compartments) to levels as high as 400–500 millimolar. In many cases these high concentrations are maintained, or they gradually develop, during exposures to adverse conditions. In some species, however, notably the euryhaline crab-eating frog (*Rana cancrivora*) of southeast Asia, the high concentrations are essentially permanent during the adult life of the animal.

This tolerance of very high urea levels offers the organism a variety of benefits, but it also has necessitated some significant additional adaptations. As an adaptation to high external salinities, tolerance of high urea significantly aids in the maintenance of internal hyperosmoticity, thus eliminating problems of water deficits. As an adaptation to extended torpidity, this tolerance serves simultaneously to lower the vapor pressures of body fluids, thus reducing rates of evaporation, and to balance changes in soil water potentials so that soil water can be absorbed cutaneously for as long as possible (Boutilier et al. 1992; Shoemaker et al. 1992; Pinder et al. 1992).

Significant additional adaptations for torpidity include several adjustments that can be made in kidney and bladder function, including continuing production of maximal urine concentrations (meaning isosmotic with the blood), reductions in glomerular filtration rates, and high levels of water and solute reabsorption and recycling from the bladder. Other adaptations are substantial changes in patterns and mechanisms of intracellular osmotic and ionic regulation in all body tissues. Among the more striking intracellular changes are shifts in the levels of accumulation of several types of free amino acids (Boutilier et al. 1992).

Recent important papers covering these and many related matters are: (for amphibious fishes) Davenport and Sayer 1986, Horn et al. 1986, Evans 1987, Iwata 1988, and Ip et al. 1990; (for amphibians) Wood et al. 1989, Boutilier et al. 1992, Shoemaker et al. 1992, and Pinder et al. 1992.

The degree to which the detailed biochemical mechanisms for ammonia production and urea synthesis are the same in fishes and amphibians is now a matter of substantial debate. There also is considerable discussion concerning the evolutionary origins of all the associated processes and of the multiple functional roles played by urea in different situations. Some important recent references in these debates are: Mommsen and Walsh 1989, Wood et al. 1989, Griffith 1991, Mommsen and Walsh 1991, and Atkinson 1992.

Nevertheless, the evidence is strong that all major features of waste nitrogen metabolism and excretion that occur in most amphibians also occur in many completely aquatic fishes and in many amphibious fishes. Thus physiological processes and mechanisms for handling ammonia and urea appear to be preadaptations for terrestrial life that first evolved in the fishes. The perfection of methods for synthesizing and excreting large amounts of uric acid and urate salts may be a post-transition development.

9.6 Sensory Biology

The magnitude, variety, nature, and scope of the changes in sensory systems that have accompanied the vertebrate water-land transition are all dramatic and spectacular. Unsurprisingly, they have attracted and continue to attract large amounts of research interest. The rate of publication of both research papers and book-length treatments of topics significantly related to the focus of this book is high. Important books published since 1985 include: Ali and Klyne 1985, Wever 1985, Bullock and Heiligenberg 1986, Atema et al. 1987, Ewert and Arbib 1987, Roth 1987, Fritzsch et al. 1988, Coombs et al. 1989, and Douglas and Djamgoz 1990. A few selected recent review articles are: Land 1987, Fritzsch 1991, Goldsmith 1990, Narins 1990, Lombard 1991, Wilczynski 1992, Burggren and Just 1992, and Land and Fernald 1992. Given the diversity and complexity of the relevant sensory adaptations, and the similar features of the literature just cited, we restrict ourselves here to mentioning a few aspects that appeal to us as particularly interesting.

Although tetrapod auditory, visual, and olfactory systems differ in many important ways from the same systems in fishes, including coelacanths, there are some important anatomical resemblances between the groups that probably imply significant functional similarities as well. One example is the inner ear of *Latimeria*, which resembles in structure, position, and innervation pattern the basilar papilla of the inner ears of amniote tetrapods. No similar structure exists in lungfishes (Fritzsch 1987). Another similarity between fishes and amphibians is seen in structural features of the telencephalon of the brain of the goldfish (*Carassius auratus*), which provide support for the idea that sensory tracts projecting from the thalamus to the telencephalon are patterned in the goldfish in ways remarkably similar to patterns found in terrestrial vertebrates. Echteler and Saidel (1981) suggest that, despite differences in both embryogenesis and cellular organization, the connections and possibly the functional organization of the forebrains of teleosts and land vertebrates are closer than had been expected.

Amphibious fishes, like other bony fishes, possess well-developed lateral line systems. We know of no specific studies of lateral line function in any amphibious fish, but it is unlikely that their primary operations differ significantly from those of completely aquatic fishes—i.e., they are the main detectors of mechanical vibration stimuli in the environment while the fishes are in water. Whether or not the lateral lines of amphibious fishes function while the animals are out of water is unknown. All three orders of amphibians also have functional lateral lines. These sense organs are present in aquatic larvae, adult pipid frogs, caecilians, aquatic adult salamanders, and adult salamandrids that are either aquatic after a terrestrial stage or remain aquatic. Lateral lines are lost by most amphibians during metamorphosis, and they seem never to develop in forms with direct development. In those post-metamorphic, terrestrial salamanders retaining neuromast organs lateral lines are covered by epidermis and probably do not function.

Larval caecilians and aquatic salamanders also have ampullary organs (that is, specialized small sensillae) on their heads that are fewer in numbers than the lateral line neuromasts, but these sensillae are often arranged in rows parallel to rows of neuromasts. Such organs have been shown to be electroreceptors similar in function to those of elasmobranch and various freshwater electric fishes. No electrosensory system is known to exist in the amphibious fishes, but structurally similar organs occur in the living lungfishes. The rostral organ of *Latimeria* may also function in this way (Wilczynski 1992; Burggren and Just 1992).

Neither the mechanoreceptive lateral line system nor the electroreceptive ampullary system survive the transition from water to land. Possible reasons for this are speculative (Fritzsch 1990). However, to our knowledge all vertebrate sensory systems underwent some sort of change in the course of the transition.

The auditory systems of both fishes and amphibians are perennial favorite subjects for students of the vertebrate water-land transition. Many books have been devoted to these systems. We note here only that there is no evidence that living amphibious fishes use sound production and audition for communication with each other. Amphibians, of course, especially many anurans, use sound for many purposes and have evolved highly complex behavior patterns around its use.

Many other illustrations could be added to this list, involving each of the external sensory systems. A possibility that emerges from this discussion is that the dramatic changes that occurred in all vertebrate exteroceptive systems as a result of the water-land transition may have been associated with a central nervous system that did not change substantially in many ways, but was broadly preadapted for terrestrial life both structurally and functionally. The many issues raised by this possibility deserve further research attention.

9.7 Reproduction and Development

Reproductive mechanisms and processes at the organismic level are little studied and poorly known for the amphibious fishes. The fact that *Latimeria* is ovoviviparous is interesting but not necessarily relevant. Reproduction has, however, been widely and intensively studied in amphibians. Twenty-nine different reproductive modes are known to occur in the anurans alone. Reviews including some information about amphibious fishes are Breder and Rosen 1966, Potts and Wootton 1984, Munro et al. 1990, and Sayer and Davenport 1991. Recent major reviews of amphibian reproduction are Hanke 1990, Jorgensen 1992, and Burggren and Just 1992. Norris and Jones (1987) provide detailed reviews of most aspects of hormonal controls on reproduction in both fishes and amphibians.

Important recent papers on physiological aspects of the development of amphibians were produced by Bradford (1984, 1990), Bradford and Seymour (1985, 1988a, 1988b), Burggren et al. (1990), del Pino (1989), Dupré and Petranka (1985), Galton (1988), Geiser and Seymour (1989), Hanken (1989), Seymour and Roberts (1991), Townsend and Stewart (1985), and Uchiyama et al. (1990). Review papers on metamorphosis were produced by Alford and Harris (1988), Dent (1988), Malvin and Heisler (1988), Semlitsch and Wilbur (1988), Shaffer and Breden (1989), and Fritzsch (1990). In this section, we present only a few of the points of greatest relevance to the water-land transition.

9.7.1 Aquatic Fishes

Completely aquatic fishes that move onto shore to spawn and for which egg development takes place more or less terrestrially could arguably represent one of the earliest stages in the water-land transition. Many fish species spawn in the upper intertidal zones of beaches, rocky coasts, and estuaries. At least eight families of marine, brackish-water, or diadromous fishes include representatives that reproduce in this manner (Clupeidae, Galaxiidae, Osmeridae, Cyprinodontidae, Atherinidae, Blenniidae, Cottidae, Tetrodontidae). Spawning in many of these forms involves what are essentially breath-hold air dives for the adults. Most species synchronize their spawning activities with the times of highest tides at the appropriate seasons of the year. An hour or two before the highest high water levels occur, they may swim up onto sandy or gravel beaches and, between waves, deposit their eggs within the sediments. Or they may swim into masses of seaweed on rocky shores or vegetation in salt marshes and deposit their eggs on the plants or inside hollow objects (e.g., empty pelecypod shells). The eggs then develop, but suspend hatching until the next extreme high tide, often two to four weeks later. When the water returns, they rapidly hatch and the larvae swim away (Potts and Wootton 1984; Sayer and Davenport 1991).

Comparable reproductive behaviors occur in several species of freshwater neotropical characins (family Characidae). The few species studied either lay their eggs on leaves hanging low above stream surfaces or spawn in groups in moist areas along the shores of streams. The eggs of these fishes hatch fairly quickly and the larvae drop into the water or wriggle their way through surface films of water to the nearby water bodies.

A major benefit of all these spawning maneuvers is to separate the fertilized eggs from aquatic egg predators, which often include the adults of the same species. This strategy is not, however, without risks. A major hazard is desiccation of the eggs (Potts and Wootton 1984; Sayer and Davenport 1991).

It is easy to visualize possible evolutionary scenarios that could follow from these shoreline reproductive patterns. Larval stages could be selected for that would gradually become resistant to desiccation and more mobile out of water, and that would develop more rapidly so as to be able to feed effectively on land. The process could then continue to eventually produce successfully amphibious organisms.

9.7.2 Amphibious Fishes

Even though reproductive strategies in living amphibious fishes have not been studied intensively, observations have been made about aspects of breeding in several groups. The best known are the mudskippers (family Gobiidae) and the rockskipper blennies (family Blenniidae).

The mudskippers are small to medium-size fishes (adults usually 5–20 cm total length) that are often abundant residents of mudflats in mangrove areas in the tropical and subtropical Old World. There is substantial diversity within the group in many respects, but as far as is known all species spawn and care for their eggs in small chambers they make at the lower ends of burrows that they dig in the mud. These burrows are usually entirely or almost entirely water filled, even during periods of low tides. Since most mangrove mudflats are composed of very fine sediments and contain large populations of microorganisms and substantial amounts of organic material, the interstitial waters of these flats are usually strongly hypoxic or completely anoxic. Water samples taken from within active mudskipper burrows indicate the same conditions there (Gordon et al. 1978). Thus mudskipper spawning, egg development, and hatching all appear to take place under conditions of significant restrictions on oxygen supply. Nothing is known about how this is possible. Very small, barely postlarval mudskippers can often be observed hopping about on mudflat surfaces during breeding periods (Brillet 1986; Sayer and Davenport 1991).

As their names imply, rockskipper blennies live primarily in rocky coastal environments. They are relatively small (adults about 10 cm total length); as juveniles and adults, they are just as terrestrial as are mudskippers. Reproductive activity has been observed in a few species of rockskippers from the western Indian Ocean and Red Sea. These animals have fairly elaborate courtship behaviors that are similar to those of related aquatic blenny species, and that result in egg deposition in rock crevices and holes like those used by their aquatic relatives. However, courtship, egg laying, fertilization, and development all take place out of water, at nest locations in the upper intertidal zone. Nest holes often slope upward, restricting water entrance. Embryonic development therefore takes place with the eggs mostly if not completely out of water. Nothing is known about just when the eggs hatch and how the larval fishes return to water, but the presumption is that hatching occurs during high tide periods or when wave wash reaches nest locations (Abel 1973; Brillet 1986).

Thus two comparably terrestrial groups of amphibious fishes, both living in the tropics and subtropics, have dramatically different reproductive patterns and sites. Both strategies are obviously successful; the fish populations in suitable habitats often number in the thousands and at high densities.

It is also important to note that the climatic regimes in the onshore areas adjacent to the shores where these fishes live are varied, ranging from humid tropical forests to austere deserts. Thus these different reproductive patterns function well in many different ecological contexts. Comparable situations exist with respect to reproduction in the many types of temperate zone and subarctic amphibious fishes. Overall, we conclude that no serious constraints based on reproductive systems would have limited the types of shores that the early vertebrates might have crossed during the evolutionary transitions from water to land.

9.7.3 Amphibians

The diversity of amphibian reproductive patterns is so great that it would be arbitrary to single out one or even a small number for special discussion with the implication that they are more relevant to our subject than are the others. Duellman and Trueb (1986) presented a detailed review of all important aspects of what they term amphibian reproductive strategies. To provide a sense of how varied the amphibians are, we outline here some of the basic facts relating to only one aspect—reproductive mode.

The three orders of amphibians are strikingly different from one another in reproductive mode. Caecilians (the order Gymnophiona) apparently all use internal fertilization, and about 75 percent of caecilian species bear living young. The remaining species are oviparous, laying their eggs either in water or on land. Aquatic egg layers all have aquatic larvae, while terrestrial egg layers may have aquatic larvae or direct development. The viviparous majority may or may not have aquatic larvae. The inferred evolutionary sequence is from primitive aquatic oviparity with aquatic larvae to increased terrestriality with oviparity, then direct development, and finally viviparity. Viviparous aquatic forms with aquatic larvae are considered divergent.

Salamanders (the order Caudata) use seven different reproductive modes. The basic distinction is between forms using external fertilization, and having aquatic eggs and larvae (about 10 percent of species), and the 90 percent of species using one of the six subtypes starting with internal fertilization. The inferred evolutionary sequence among the forms using internal fertilization is: (a) eggs and larvae aquatic; (b) eggs terrestrial, larvae aquatic; (c) eggs and larvae both terrestrial, but larvae do not feed; (d) eggs terrestrial, development direct; (e) eggs retained in the oviduct of the female, developing and hatching there (ovoviviparity); and (f) eggs retained in the oviduct and developing there into fully developed juveniles (viviparity).

Only a few species of frogs (order Anura) have internal fertilization; well over 90 percent of forms have external fertilization. Three generic reproductive modes for anurans are recognized based on sites of egg development: (a) aquatic eggs; (b) terrestrial or arboreal eggs; and (c) eggs retained in oviducts. A total of twenty-nine specific modes are recognized within these categories: eleven variations on aquatic development; sixteen variations on terrestrial or arboreal development; and two variations on egg retention (ovoviviparity and viviparity). The range of patterns extends from eggs laid and tadpoles that feed in water through eggs being carried about by one of the adult parents and developing directly into froglets. Direct development occurs in several different sets of circumstances.

There is no way of knowing if additional reproductive modes might have existed in earlier geological time periods. Whether or not they did, the extant diversity certainly demonstrates that amphibians can successfully reproduce and develop under an amazing array of environmental circumstances. There is further diversity within many of the modes listed, in terms of such aspects as durations of larval periods (ranging from days to years), the occurrence of neoteny, and resistances of different developmental stages to difficult or severe environmental conditions. What is more, amphibians are masters at exploiting available microhabitats and/or microclimates. The final result is that amphibians occur in almost all terrestrial habitats outside of truly polar environments and active volcanoes. Once again there are few conceptual constraints on possible paleoenvironments for the vertebrate water-land transition.

9.8 Behavior and Ecology

Each of the preceding sections in this chapter presented significant information relating to the topics of behavior and ecology. None of this information will be repeated in the discussion that follows.

9.8.1 Amphibious Fishes

Direct observations of limited aspects of the behavior and ecology of the deepwater coelacanth, *Latimeria*, have been made in recent years through the use of submersibles in the Comoro Islands. Results published so far do not appear relevant to the concerns of this book (Fricke et al. 1991). Behavioral and ecological information on expressly amphibious fishes is available, but widely scattered. Gibson (1982, 1986, 1988) and Sayer and Davenport (1991) discuss aspects of the subject. Additional information is provided by Gordon et al. (1985), Brillet (1986), and Brown et al. (1991).

Individual and group social behaviors of great complexity are characteristic of terrestrial aggregations of many species of subtropical and tropical amphibious fishes, especially the mudskippers and rockskippers. These behaviors relate to several things: courtship, reproduction, social status within groups (pecking orders), and territoriality. There is considerable variability in these respects both within species (behaviors of populations in different geographic regions are often different) and between species.

A particularly striking example of a complex behavior pattern significant in several contexts simultaneously is burrow construction by the most terrestrial species of mudskippers, the members of the *Periophthalmus koelreuteri* group. The surface openings of the burrows of these fishes vary in form from simple holes in the surfaces of mudflats to mud towers or castles of considerable size in comparison to the fishes that make them. The burrows are made by the fishes pressing their mouths into the mud and carrying bite-size portions elsewhere. The forms that make simple burrow openings drop the mud on the surface of the flat some distance from their burrows. The castle builders create circular structures a few times larger in diameter than the burrows themselves. When they have completed the structures, the castle builders use the parapets as resting places from which they survey their surroundings for food, danger, and their conspecifics. These species are more likely than the simple burrow builders to be territorial. When approached by conspecifics the castle builders will often participate in aggressive displays involving color changes, erection of vertical fins, head-bobbing, maneuvering for physical position and, sometimes, active fighting (Brillet 1976, 1986).

In contrast to the mudskippers, the rockskippers cannot make burrows in the rocks on which they live. They show no territoriality unless they are males defending a crevice or hole containing eggs. Rockskippers are primarily herbivorous, while the mudskippers just discussed are carnivores or omnivores. The rockskippers show much less aggression among themselves than do the mudskippers. They will often graze quietly on algal mats in tightly packed groups, frequently in contact with one another or climbing over one another. Only infrequently and briefly will they bob their heads or raise their vertical fins toward another fish. Their body colors do not change much in the course of most social interactions, and they almost never bite each other. Overall they appear to be substantially less aggressive and hierarchical than are mudskippers (Brillet 1986; Brown et al. 1991).

The reason for describing these behaviors in some detail is to illustrate both the levels of behavioral complexity that exist in some amphibious fishes and some of the ecological interactions associated with these behaviors. It seems probable that the castles and the territoriality of castle-building mudskippers serve to limit population densities of these fishes. Since they are largely carnivorous, their food supply is more limited or unpredictable than is the algal supply used by the rockskippers, and density restrictions help ensure adequate supplies for the fishes that are there. A problem with this adaptational account of behavioral differences between rockskippers and mudskippers is the lack of territoriality in the mudskipper builders of simple burrows; these burrowers live at no higher population densities than do the castle builders and in what appear qualitatively to be similar habitats. The forms with simple burrows also have many aggressive interactions with their surrounding conspecifics, but these conflicts are not over territory.

There are ways different from and in addition to reproductive strategies in which these two groups of amphibious fishes succeed in their respective environments by means of different adaptive scenarios. The behavioral complexities, overall, are comparable with those involved in aggregations of amphibians. Indeed, amphibious fish behavior may actually be substantially more complex, since the fish societies described exist year round while most amphibians aggregate only for short periods for courtship and mating. Temperate zone and subarctic species of amphibious fishes are not known to have social interactions at all in contexts other than spawning. Possible bases for the differences between amphibians and amphibious fishes are unknown. Once again the specific behaviors of the amphibious fishes ancestral to the amphibian lineage or lineages must remain a mystery, but conceivably that behavior was already varied and complex.

There is also substantial diversity in the environmental conditions that prompt amphibious fishes that do not live in the marine intertidal zone to demonstrate their terrestrial capacities (Sayer and Davenport 1991). The usually nocturnal overland excursions—often over long distances—that are made by European eels (*Anguilla anguilla*) moving from one body of water to another are near legendary. Nothing substantial is known about the nature of the stimuli leading to these movements, or about how the eels "know" where they are going and how to get there. In the mountains of south Asia, including the Himalayas, several species of torrent-dwelling catfishes have been anecdotally described as climbing the damp rock walls behind high waterfalls. The spontaneous short excursions onto land made under laboratory conditions by *Erpetoichthys*, a freshwater polypterid fish of Africa, were mentioned earlier in this chapter.

The proximate motivations for these overland excursions are of five possible types: foraging for food; moving to better quality habitats; migrating; trying to remove or eliminate ectoparasites; and escaping possible predators. Here again there is ample room for a diversity of plausible evolutionary scenarios that might have led to the origins of the Amphibia (Sayer and Davenport 1991).

9.8.2 Amphibians

Substantial portions of at least eight chapters in Duellman and Trueb 1986, and of almost all chapters in Feder and Burggren 1992, are devoted to amphibian behavior and ecology. Our interest here is in presenting the range of environmental situations in which the living Amphibia occur.

Geographically amphibians occur on all continents except Antarctica and on many oceanic islands. The altitudinal range of the group extends from below sea level (species living in desert areas that are below sea level in North America) to about 4,000 meters in several major mountain ranges. Amphibians occur from the tropics to the southern tips of Africa and South America, and they can be found to the southern edge of the subarctic in North America and Eurasia. They live in most varieties of freshwater habitats, excepting hot springs and highly saline inland waters. They also live in severe deserts and in the marine mangrove forests of southeast Asia. Some are arboreal while others live in caves and underground bodies of water. Many forms are lifelong burrowers; some seldom emerge from the ground. As Duellman and Trueb comment: "In the successful attainment of independence from water and colonization of land, amphibians have undergone a remarkable adaptive radiation, and the living groups exhibit a greater diversity of modes of life history than any other group of vertebrates" (1986:1).

9.9 Summary

The reviews just completed demonstrate and indicate several things. First, the living amphibious fishes and Lissamphibia successfully occupy a variety of highly diversified and geographically widely distributed habitats. Second, each group has achieved its success through the use of many combinations and permutations of both structural and functional features at multiple levels of organizational complexity. Third, applying the uniformitarian principle (set forth in chapter 1 and employed in the final chapter) to this data base leads us to conclude that the early stages of vertebrate transitions from water to land might well have occurred under any or all of a wide variety of environmental circumstances—certainly not only in the lowland humid tropics. We also conclude that the extinct presumptive ancestors of amphibious rhipidistian fishes could well have been both morphologically and functionally diverse. This scenario would be consistent with the morphological diversity of the few known fossils of ichthyostegalian amphibians.

The living coelacanth, *Latimeria*, provides some intriguing insights into what the ancestral crossopterygians might have been like, yet it can also lead biologists seriously astray. The coelacanths have had a separate evolutionary history for more than 360 million years. *Latimeria* itself has been a mesobenthic marine genus for some substantial, though probably much shorter, period of time. There is no reliable way to determine how much, or which parts, of the story told by *Latimeria* relates to its ultimate ancestry and how much, or which parts, relate to its own history.

The living amphibious fishes as a group probably do not now represent incipient series of possible future vertebrate invasions of the land. There are no known data determining the evolutionary ages of any of these taxa; but, with the possible exception of the polypterid *Erpetoichthys*, it seems likely that most kinds of extant amphibious fishes are geologically relatively young. Two major sets of factors would seem likely to prevent these extant genera from going on to evolve greater terrestriality. First, they all are still bony fishes, using the basic body plans and locomotory mechanisms associated with that state. These structural and functional morphological conditions appear not to be as utilitarian and operationally advantageous in terrestrial contexts as is the tetrapod plan and patterns of locomotion. Second, all major potentially suitable terrestrial niches and habitats are surely already occupied by tetrapods. Unless the tetrapods are eliminated, and without associated widespread environmental destruction, the amphibious fishes are likely to be continuously out-competed in that realm.

A surprisingly large proportion and variety of the basic vertebrate features that are considered part of the terrestrial mode of life appear to actually have been preadaptations first evolved among the fishes, presumably for quite different reasons. These features include: (a) major features of the organization and programming of those parts of the central nervous system associated with locomotory movements and sensory physiology; (b) the structure of the limbs and the basic anatomical arrangements of musculoskeletal elements associated with tetrapod postures and locomotion; (c) most of the basic metabolic and physiological properties associated with aerial respiration, water and solute regulation, metabolism, and waste nitrogen production and excretion; (d) many, if not most, of the general features of reproduction and embryonic development used by the Amphibia; and (e) significant fractions of the behavioral capacities associated with terrestrial life.

At least two sensory systems important to the fishes and aquatic amphibians (including aquatic stages of forms metamorphosing into fully terrestrial adults) have not been retained in lineages that have adapted to subaerial environments. These are the mechanosensory lateral line system and the ampullary electroreceptive system. These two sensory systems seem to be the only major deletions or losses in vertebrates associated with the water-land transition.

Finally, as was true for the molluscs, leeches, and crustaceans, there are huge gaps remaining in our understanding of the who, what, when, where, and how of the vertebrate portions of the water-land transition. Much interesting, significant, and useful research remains to be done in all of these areas. We also suspect that the answer to the "why" question will long remain enigmatic.

10

Summary and Synthesis

MALCOLM S. GORDON, EVERETT C. OLSON, AND DAVID J. CHAPMAN

After eight chapters that review the facts and interpretations of the vascular plant and metazoan transitions to land, what are the key conclusions? And which of these are new and suggestive of important insights?

1. Many new finds in the fossil record have come from recent work done in at least three different modes: reevaluations of fossils already present in collections; new discoveries in the field; and the application of new approaches and techniques, especially those directed toward microfossils in ancient rocks. Collectively these finds have produced important changes in perceptions of chronologies, sequences, relationships, environments, ecologies, and geographies.

2. Entirely new lines of evidence and new perspectives are developing from the use of modern molecular biological and biochemical methods applied to living organisms.

3. New interpretations of phylogenetic relationships of organisms come from cladistic analysis of both older and newer data sets.

4. New information about the diversity and scope of functional adaptations of living organisms, especially amphibious or semiterres-

trial forms, is producing increased awareness of the great range of environments, habitats, and functions that could have been involved in the actual historic transitions. Narrow interpretations based solely on very distantly or speculatively related fossils are untenable.

In chapter 1 we outlined the theoretical framework that we have come to believe is the most realistic and useful for analyzing the evidence related to the water-land transition. We restate that framework in section 10.3.1. Based on this framework we have developed nontraditional perspectives on what the evidence tells us about how the actual transitions to land may have taken place. We present these in sections 10.3.2 and 10.3.3.

The review chapters clearly show that the invasions of the land carried out by each of the groups that successfully made that transition were complex—at least in the groups for which evidence is substantial. This complexity probably also applies to those major groups for which there is almost no evidence. An understanding of the complexities is muddled by the problems associated with interpreting the fossil record (chapters 3 and 4). We believe it has been further muddled historically by premature acceptance of specific evolutionary hypotheses that are plausible and appealing, but neither complete nor necessarily the most realistic interpretations of the available data.

Having said this, we nevertheless believe that there are many factually reliable conclusions that can be drawn about the invasions. We present these conclusions in the next two sections. We then present an overall synthesis of our views of the current state of knowledge in this area of biology, including new theoretical and conceptual interpretations.

10.1 Conclusions Drawn from the Fossil Record

The actual time intervals during which the invasions of the land occurred are known approximately for only two groups: the vascular plants and the chelicerate arachnid scorpions. Fossils of possible transitional forms have, however, only been identified for the scorpions. The best available evidence provides the following approximate chronologies for all of the major groups (most figures are minimum estimates of the ages of first occurrence of putatively terrestrial organisms).

Vascular Plants The invasion began prior to 450 mya (late Ordovician). The full range of adaptive features needed to establish and facilitate colonization had apparently evolved by about 420 mya (late Silurian). A varied and thriving land flora was established by 390 mya (early Devonian).

Prosobranch Gastropods There was no single invasion. Different families made the transition at scattered intervals over long periods of time. The oldest known terrestrial prosobranchs evolved about 300 mya (the later stages of the late Carboniferous). The second oldest families evolved about 100 mya (Cretaceous). The youngest evolved about 50 mya (Tertiary).

Pulmonate Gastropods Again there were multiple invasions at scattered intervals over long periods of time. The oldest order known evolved almost contemporaneously with the first terrestrial prosobranchs about 300 mya. The second oldest order evolved about 150 mya (late Jurassic to early Cretaceous). The third major group of living pulmonates has no known fossil record. Substantial secondary reinvasion of aquatic habitats apparently has occurred, all at unknown times.

Land Leeches The land leeches have no known fossil record. Indirect evidence indicates that origins in the early Mesozoic are possible.

Chelicerates The earliest known terrestrial arachnids evolved about 410 mya (late Silurian). The earliest known scorpions also date from the Silurian (about 420 mya), but these arthropods—including the few additional fossil scorpions known until the late Carboniferous—were all aquatic. Fossil scorpions that appear to be transitional from water to land are known from Carboniferous deposits. By the late Carboniferous (about 300 mya) the scorpions had become entirely terrestrial—as they are today, except for a few secondary reinvaders of aquatic habitats.

Isopod Crustaceans No fossils are known that are relevant to water-land transitions of the isopods. A few completely terrestrial families have limited and fragmentary records dating back to the Eocene or Miocene (50–15 mya).

Amphipod Crustaceans No fossils are known relating to the single living family of amphipods that includes terrestrial representatives.

Decapod Crustaceans No fossils are known relating to the living amphibious or terrestrial families.

Uniramians No fossils are known that are relevant to primary water-land transitions. The earliest known possibly terrestrial myriapods date from about 410 mya (late Silurian). The earliest known collembolan (apterygote hexapod) dates from the early Devonian (about 380 mya). A diverse fauna of pterygote hexapods first appears in the late Carboniferous (about 300 mya).

Vertebrates No fossils are known that specifically relate to living genera of amphibious fishes. Various of the larger groups of fishes, including some families that contain amphibious forms, do have fossil records of varying extents. No fossils are known of possible amphibious rhipidistians or protoamphibians. The earliest known amphibians date from the late Devonian (about 350 mya). The earliest known ancestral Lissamphibia date from the Triassic (about 200 mya).

The only groups for which it is possible to say anything that may be considered reliable with respect to probable aquatic ancestors are the vascular plants, the scorpions, and the vertebrates. Today's land plants are related most directly to the green algal class Charophyceae, more specifically to *Coleochaete*-like forms in the order Coleochaetales.

Until recently the evidence in support of a charophycean ancestry was derived principally from comparative biochemistry and ultrastructural details of flagellar structure and mitotic events in living representatives. This generalized hypothesis has now received substantial support from molecular-derived data, viz., amino acid sequences of proteins and nucleotide sequences of the small subunit ribosomal RNA and 5S rRNA. However, all these arguments are helped by the fact that it is very difficult to make a plausible case for an ancestry from any other group of algae that is extant today.

The case for a *Coleochaete*-like representative is based as much on default as anything else. Genera in the Charophyceae other than

Coleochaete are even further removed morphologically and ontogenetically from what the earliest land plant is believed to have resembled. There is a wide gulf between the presumptive ancestors (likely modern and fossil candidates) and the postulated earliest terrestrial inhabitants.

The traditional approach in paleobotany of looking for the key intermediate, or "silver bullet," has failed to narrow the gap. Did the aquatic ancestor belong to an algal group for which there are no modern relatives? Did the transition occur more than once either temporally or geographically? How many migratory attempts finally failed, leaving no modern relatives? There has been a reluctance to address these provocative and controversial questions. This is perhaps because traditional scientific disciplines and their research techniques have not lent themselves to obtaining the answers.

Terrestrial scorpions were apparently descended from aquatic scorpions, but the record is scanty. Living amphibians and higher land vertebrates are believed to have derived from the rhipidistian branch of crossopterygian fishes. Recent cladistic analyses of both molecular and morphological data support the position that the amphibians are monophyletic in origin. However, significant uncertainty about this remains.

There are no known aquatic antecedents for any of the other groups of metazoans.

For many of the groups for which aquatic antecedents are unknown, there *is* substantial consensus among workers familiar with them as to the general nature of the probable environments in which the transitions took place. The current consensus positions are: (i) plants: freshwater; (ii) prosobranch gastropods: some marine, others from brackish or, more probably, freshwater environments; (iii) pulmonate gastropods: freshwater; (iv) leeches: unknown; (v) chelicerates: scorpions probably were marine; (vi) isopod crustaceans: possibly marine; (vii) amphipod crustaceans: marine and brackish waters; (viii) decapod crustaceans: some from each of the three possibilities; (ix) uniramians: possibly freshwater for some, unknown for most; (x) vertebrates: freshwater.

Basing our own view on an assumption that multiple geographic origins and polyphyly were to have been expected, plus the possible mosaic nature of the record in each case, we believe that the best that can be said about the origins of these terrestrial groups is that the present statistical sampling of relevant paleoenvironments is inadequate to permit definitive statements. We note that the bases for most of these inferences are indirect lines of evidence deriving from studies of living forms.

Some significant part of the difficulty that researchers studying fossil molluscs, arthropods, and fishes have in identifying possible amphibious forms, and in defining more precisely when terrestrial forms first appear in the record, derives from the fact that the gross anatomies of the fossilized body parts of these groups do not change detectably across the water-land transition. In contrast, it is easy to tell a fossil fish from an amphibian.

10.2 Conclusions Drawn from Living Plants and Animals

Chapters 3, 7, 8, and 9 each have summaries covering many of the important points discussed. We will not here repeat the details for each group. Rather, we consider topics that relate to more than one of the major groups.

First, it is important to note that each of the phylogenetic transitions involved to some degree all or almost all of the major organ systems in the lineage making the transition. In the plants important changes took place, for example, in the chemical compositions and structures of cuticles and cell walls, in the structure and function of systems for water and nutrient transport, in reproductive mechanisms and processes, and in growth patterns and forms.

It is most unlikely that these new adaptations and changes all appeared simultaneously in one colonizing candidate. Thus the sequence of these events is an integral part in the establishment of the scenario of the transition. This is particularly important for any intermediate forms living a possible dual existence in both aquatic and terrestrial habitats.

In chapter 3 we outlined a possible order of events for the evolution of vascular land plants. It is, however, tied to an incomplete fossil record and to perceptions, based on the physiology of modern plants, of relative importance and need. The comparative biochemistry of all green algae, higher plants, and bryophytes also indicate that a number of other biochemical changes occurred with the groups involved in the transition. Plausible explanations or hypotheses for these changes have yet to be proposed. Were they a necessary adaptation for the migration, or were they a result of the transition?

In the animal groups all important systems were involved. Probable limited exceptions were the completely internal parts of some digestive systems and many of the coordination and control portions of central nervous systems. The relative amounts of change that occurred in nervous systems varied widely among the groups making the transition. For example, the sensory nervous systems and locomotor systems of the gastropods and arthropods changed relatively little, while these same systems changed substantially in the vertebrates.

Important changes also occurred at levels other than the macroscopic. Significant biochemical changes occurred in many lineages, particularly in terms of both the development of new metabolic pathways and the relative rates of usage of preexisting alternative pathways.

In addition to these and other systemic changes in plants and animals, transitions to the land in all groups involved the evolutionary development of increased structural diversity and complexity, and of greater functional versatility and adaptability. There was little loss of either structural or functional elements that the pioneering ancestors must have brought with them. The only substantial examples of losses were the elimination of the lateral line and ampullary sensory systems in terrestrial vertebrates.

Important components of the changes in all groups were the conservation and redirection in new operating modes of structures and functions originally evolved in the aquatic environment. These changes could thus be considered preadaptations. Strikingly large fractions of the properties most people would consider to be essential aspects of terrestrial adaptation in each group belong in this category.

At the same time, the diversity of new features evolved for coping with the many environmental differences between aquatic and terrestrial habitats was often large. Here again, the balances were substantially different in different groups (e.g., the sensory systems of molluscs and crustaceans were relatively little changed, while those of vertebrates changed markedly).

Overall, the clues about the water-land transitions that can be obtained by studying the living amphibious and semiterrestrial representatives of each of the major groups has added significantly to the range of possibilities we must consider with respect to both the environmental and the functional aspects of the actual historic transitions. In the final section we make an attempt at synthesizing these different lines of evidence.

10.3 Synthesis

At least four substantial efforts have been published in the past few years that attempt to synthesize the lines of evidence and data sets relating to water-land transitions in plants and various animal groups, plus plant-animal interactions. The most detailed of these efforts was by Little (1990), who worked primarily with ecophysiological and behavioral information derived from living forms, especially animals. Selden and Edwards (1989), Shear (1991), and Gray and Shear (1992) included more information on plants and also addressed many aspects of the fossil record. All three of these reviews are, however, much shorter, less complete, and less detailed than is Little 1990. Shear (1991) focuses his discussions of animals almost entirely on the insects and tetrapods; he emphasizes ecosystem evolution rather than organismic evolution. Behrensmeyer and associates (1992) provide the most comprehensive overview of ecosystem evolution that is yet available.

Our discussion here is aimed to complement and supplement the others in ways that are based on the somewhat different theoretical and conceptual approaches we have used in this book. In the interests of clarity and reasonable brevity, we deal with only those selected issues we consider important and that distinguish our own conclusions from those of others. We do not try to be comprehensive or encyclopedic.

Much of our discussion of functional features of organisms, especially those features that are specifically physiological, fits well into the general category of what Burggren and Bemis (1990) call "feasibility analysis." By this term Burggren and Bemis mean an analysis that uses functional studies of living organisms, irrespective of their phylogenetic relationships to the organisms that historically carried out the evolutionary transformations of interest, to demonstrate the real world feasibility of particular adaptive scenarios. We have called these approaches "envelopes of possibilities" (chapter 1). They might also be called applications of functional uniformitarianism.

10.3.1 Central Assumptions

Scientific method in phylogenetic and evolutionary biology strongly emphasizes parsimony in the generation of hypotheses. We fully agree with the position. However, the results of our factual overview of water-land transitions lead us to suggest that, at least for these processes, several factors beyond those traditionally considered should be added to the data bases from which evolutionary hypotheses derive. The addition of these factors significantly changes one's perspective on the issue and leads to different hypotheses. We consider the new hypotheses, in fact, to be more parsimonious than are the traditional ones with respect to important features of the fossil record and an array of environmental, biogeographic, and functional adaptational matters relating to organisms.

Our approach is somewhat similar to that used by Gans (1989) in his review of the evolutionary origins of the vertebrates. Gans, however, dealt with a phylogenetic transition rather than an ecological one.

Our starting point is the set of four central assumptions presented in chapter 1. We recapitulate them here for convenience, with some rephrasing and further elaboration:

1. Diversity of Paleoenvironments and of Ancestral Organisms The principle of uniformitarianism leads us to believe that, at least since the Cambrian, paleoenvironments potentially suitable for water-land transitions were not qualitatively different from comparable environments today. Paleoenvironments were just as structurally diverse and spatially and temporally variable. They were also comparably large in extent, and comparably distributed along the edges of and, for freshwater, within the paleocontinents—however those continents were spatially distributed. In addition, the ancestral organisms of the future terrestrial colonists were also qualitatively as structurally and functionally diverse and adaptable as are living amphibious and semiterrestrial representatives of the same major groups.

2. Polyphyletic Origins The probability that there were single transitions within any major group is vanishingly small, if not zero. Bases for this include the considerations outlined in the previous section. Then too, and excepting possibly the plants (for reasons explained in chapter 3), organisms likely to have been ancestral in water-land transitions probably had relatively limited physical mobilities throughout their entire life histories. Relatively limited dispersal capacities, in turn, probably often resulted in the complete genetic isolation of widely dispersed populations of pioneering taxa. More fundamentally, multiple taxa within each major group, many of which could have been widely dispersed, could have been the sources of the migrants, with those different taxa also genetically isolated from one another. Finally, there were the inevitably different selective situations that existed for different populations within single taxa as well as for different taxa living on different, widely separated continents or in far-flung locations along the borders of or within giant continents. The overall results of these circumstances were, most probably, multiple sets of transitions with resultant organisms generally resembling one another even though the organisms were widely dispersed geographically and strongly polyphyletic. The resemblances could be due to parallelisms resulting from preexisting similarities between the organisms or convergences.

3. Incompleteness of the Fossil Record For many groups it is probable that the ancestral forms remain undiscovered or have vanished without record. This could be particularly true for small plants or animals lacking hard parts or readily preservable tissue or structures.

4. False Starts in the Fossil Record There is a strong probability that, at least in the present context, the fossil records of the major groups having significant records may actually be complex patchwork conglomerates of segments of multiple sequences, rather than components of single sequences. Each major group that was ultimately successful could easily have had several to many lineages that developed terrestrial adaptations part way, or even all the way, but then became extinct. We have no way of distinguishing such false starts in the fossil record from those lineages that proved successful. It is probable that present views of evolutionary "main lines" for terrestrial invasions are to some extent based on fossils that are not members of single clades.

Our factual overview leads us to conclude that there are only two major groups for which enough is known to permit the development of detailed possible scenarios. These are the vascular plants and the vertebrates.

10.3.2 Plausible Scenarios for the Evolution of Vascular Plants

The relationships between algae, bryophytes, and presumptive early vascular plants present a series of problems. These groups left a poor fossil record that is difficult to interpret. The algae themselves are polyphyletic and similar morphologies have clearly evolved more than once in unrelated groups. Phylogenetic relationships based on morphology are thus difficult to interpret. Independent evidence— e.g., paleobiochemistry—often further clouds, rather than clarifies, issues. We believe that attempts to find and interpret intermediates in the fossil record may be doomed to failure except through serendipitous luck. What one looks for is frequently determined by preconceived notions; should an intermediate in the evolutionary continuum be discovered, its identity as such may not be recognized.

Chapter 3 presented what we consider to be the most plausible scenario for the evolution of land plants. This scenario differs in many respects from other current views. It is, however, the scenario that is most compatible with the framework (notably, the four assumptions presented in the previous section) that we believe to be best for deriving and evaluating transition scenarios. Our scenario does mesh broadly with most others in that it does not envision the bryophytes as giving rise to vascular plants; vascular plants rose separately from their own algal lineage.

Chapter 3 points to a number of caveats and potential problems in this scenario. No scenario can be put forth with absolute assurance with the current state of knowledge. In reaching our conclusions, however, we have relied on data and ideas drawn from an array of different techniques and approaches: paleobotany, paleobiochemistry, comparative biochemistry, ontogeny and morphology of extant organisms, physiological requirements and constraints of extant organisms, and evolutionary proposals based on molecular data.

Overall, *we are willing to entertain an invasion of plants earlier than many would accept. We are unwilling to accept at this time a "single event" hypothesis,* even though hard evidence to support multiple temporal or geographical invasions is lacking. The protists, which include algae, are notorious for their diversity and the complexity of their evolutionary relationships. This complicates the search for the simple linear relationship one might hope for with respect to vascular plants. We are willing to consider that there were probably a number of invasion attempts that failed without establishing identifiable land plant descendants—and that some fossils of these false starts, moreover, may be among those to which paleobotanists have traditionally looked for clues.

These departures from traditional ideas represent a challenge. How might they be evaluated against the reigning views? Gray and Shear (1992:455) state: "All hypotheses about the ecology and the evolutionary history of terrestrial life must be tested by the evidence in the rocks." We believe that searching for the answer through interpretations of only the fossil record will not succeed. Working with the fossil record is only one approach of many.

10.3.3 Plausible Scenarios for the Evolution of Terrestrial Vertebrates

Our proposal for the vertebrates is significantly different in pattern from our proposal for the water-land transition of plants, largely because of the probably much more limited dispersal capacities of possible rhipidistian amphibian ancestors as compared with the spores of early plants.

We begin with the major facts concerning the fossil records of the early Amphibia. The salient points are: (a) All of the earliest known amphibian fossils (labyrinthodonts) are late Devonian in age. (b) These fossils occur in deposits laid down in relatively low latitude regions of the two continents of Laurussia and Gondwana; the specific locales were at that time separated from one another by thousands of kilometers of open ocean and land (figures 2–6 and 2–7). (c) These earliest fossil amphibians were already at that time quite large in size, structurally diverse, fairly specialized, and phylogenetically well differentiated from one another. (d) There are no plausible connections between these earliest fossil amphibians and living amphibians. That is, the morphologies of these two groups are so different that there is no scientifically justifiable way to derive the latter group from the former group.

The next known amphibian fossils are substantially younger (Carboniferous and Permian in age) and much more differentiated, diverse, and specialized. These later fossils include representatives of several additional groups of extinct forms, but none have plausible connections with living amphibians. The fossil records of the modern groups of amphibians extend back only to the early Triassic.

We believe that our proposed theoretical and conceptual framework accommodates this factual picture with fewer arbitrary assumptions required than do more traditional views of early amphibian evolution. Our framework would not require a single main line of phylogenetic relationships. *Thus the Laurussian and Gondwanan labyrinthodonts were at best only distantly related members of a grade of morphological differentiation. Some or all of the extinct groups could represent transitions that left no descendants among living tetrapods.* These groups could be quite unrelated to whatever group or groups were the actual ancestors that later led to modern forms. The actual ancestors might well remain completely unknown, or they may have already been discovered—there simply is no way to tell. In any case, the structural features of the known forms can be informative as "envelopes of possibilities" for the structures of the then contemporary ancestors.

Chapter 9 documented the great ranges of morphological and functional differentiation and the geographic and environmental diversity that occur among the living amphibious fishes and true amphibians. Adding these facts drawn from extant organisms to our framework leads us to further conclude that

1. The environments of the earliest known amphibians, while certainly important for those animals, may or may not be relevant for understanding of the initial stages of the vertebrate emergence. Those environments were apparently located at relatively low latitudes, at low elevations, and in freshwater. They may have been warm and humid. None of this necessarily implies or requires that the actual ancestral forms lived in similar situations. Based on what is known of living amphibians, transitional environments could have

included higher elevations, cool to seasonally cold temperatures, or seasonally or regionally dry to xeric conditions.

2. Again looking at the environments of living amphibious fishes, the earliest amphibious rhipidistians (yet unknown) could have been marine or estuarine or even freshwater forms. They could have lived anywhere from tropical to subarctic latitudes. Amphibious forms could have arisen from freshwater populations then living at elevations near sea level to thousands of meters high in mountainous regions. They could have lived in almost any littoral or riparian habitats, from rocky coasts to protected estuaries, rapid streams, subalpine lakes, or marshes.

3. The unknown earliest amphibious rhipidistians might have begun their emergence initially by laying their eggs out of water, near the water's edge. They could have done this in shore environments that were relatively calm, like salt marshes or mangrove habitats in bays or estuaries, or in riparian freshwater regions. They could have attached their eggs to intertidal seaweeds on rocky coasts or buried them in sandy beaches along the more disturbed coastlines of the open ocean. The first stages in terrestrial adaptation could have been taken not by the adults, but by the hatchlings that had to make their way back from land to nearby, possibly tidal, water bodies. There are many living fishes that successfully do all of these things.

4. The unknown earliest amphibious rhipidistians and their also unknown earliest amphibian descendants did not necessarily have lungs to breathe air or legs to use for locomotion. They did not necessarily have heavily scaled or armored skins or any other special morphological adaptations to permit aerial respiration, protect themselves from evaporative water loss, orient themselves adequately, and so on. All these characteristics could be results of evolutionary radiations in later stages of the transition. The first emergers could have been quite successful for extended periods of time out of water as unmodified fishes or almost fishlike amphibians. All these features can be found among the living amphibious fishes.

Overall, we conclude that, until new and different fossil protoamphibians are found that force narrower interpretations, there are few logically justified constraints for limiting hypotheses concerning the specific properties of the earliest stages of the evolution of terrestrial vertebrates. Even when such fossils are found, if ever, there will not be assurance that they represent actual ancestral forms.

10.4 Further Applications of Our Theoretical Framework

Application of our theoretical framework to other aspects of the water-land transitions leads us to some additional observations:

First, we agree with other authors that the most likely basis for the absence from the fossil record, at least so far, of the remains of most animals belonging to cryptic groups is that cryptic environments are not conducive to fossilization. This circumstance may also account for the lack of known fossils relevant to water-land transitions for the many groups we have discussed for which neither aquatic antecedents nor transitional forms are known. We also agree that cryptic environments (e.g., subsoil, rocky coastal seaweed mats, and beach and forest floor litter habitats) were the most likely routes for land invasions by such groups as land snails (both prosobranchs and pulmonates), land leeches, isopods, and amphipods. It is also possible that these habitats played a role in the emergences of some decapod crustaceans, especially if those forms first ventured from the water as newly metamorphosed small crabs. It seems unlikely that cryptic habitats were very important in the vertebrate transitions, aside from the gymnophionans.

A corollary that can be drawn from the application of our conceptual framework is that to the extent that the transitions in the major animal groups were first made via progressively longer and more effective land excursions by larger juvenile or adult life stages, the indirect evidence strongly indicates that only the most extreme environments should be considered as unlikely sites. The living amphibious and semiterrestrial molluscs, leeches, and crustaceans, in addition to the fishes and amphibians, successfully occupy a wide range of habitats in all relevant respects: geographically, altitudinally, thermally, in terms of salinity or oxygen, and so on. There is no valid reason to assume that the various ancestral groups were significantly less flexible and adaptable.

Finally, present understanding of the evolution of paleoclimates and paleoatmospheric compositions makes it improbable that global environmental constraints played an important role in determining when the land invasions by either plants or animals began. For the plants suitable terrestrial potential niches were present for long periods before the invasion started. Reasons for the delay are unknown. It is, however, possible that slow development of terrestrial soils played a role. It seems probable that the most significant constraints on the animals generally had to do with the early insufficiency of suitable ecological niches and supplies of food on the land. The evolution of land plant diversity and of substantial populations of small cryptic animals were probably major influences. It also seems probable that, while potential terrestrial niches and abundant food supplies were positive factors facilitating the beginnings of invasions some negative factors in aquatic habitats operated as well. The most important of these may well have been intense levels of predation, disease, and parasite pressures, plus competition for food resources—all of which had plenty of time to develop in aquatic habitats. In addition, the environmental stresses of drought drying up bodies of water or of other extreme conditions, such as prolonged anoxia, may also have played roles. These latter circumstances seem likely to have been most significant in relatively restricted areas and for (geologically) relatively short periods of time.

We recognize that for cladistically oriented readers our basic framework creates significant practical problems. Specifically, our approach suggests that one regard many fossil sequences as possibly being mosaic composites. And our conclusion that some higher level vertebrate taxa are likely to be polyphyletic (implying the occurrence of substantial parallelism and convergence) could require significant revisions in classifications in order to bring them in line with cladistic principles. Our view, however, is that what we propose is a strongly uniformitarian and highly probable interpretation that warrants consideration.

We believe that the theoretical framework we have used to probe the water-land transitions of vascular plants and metazoans could also be usefully applied to a range of other evolutionary questions. It could be used to partly reinterpret some of the scenarios for vertebrate origins presented by Gans (1989). The origins of other animal groups could also be viewed from the same perspective. We hope that this framework will at least have heuristic value in raising possible alternative interpretations of a range of issues pertaining to the fossil record.

We wish to close this book with a question concerning the nature of reality:

Which is more parsimonious: (a) a framework leading to flexible scenarios like those we propose, which seems a clear application of uniformitarian principle; or (b) a set of rigid cladist scenarios requiring that all descendants of particular groups, however geographically widespread they may be, derived from only single local populations of those groups? Such rigid scenarios also often require that the groups evolve over geologically short periods of time; their proponents tend to disregard what may have been the severely limited powers of dispersal of the groups, and they also pay little attention to how varied the products of the postulated evolutionary radiations may be. Perhaps the nodes of phylogenetic branching for many terrestrial groups do not lie in the earliest ancestral amphibious forms. The nodes may actually be deeper, among the aquatic ancestors of those forms.

References

Aarset, A. V. 1982. Freezing tolerance in intertidal invertebrates (a review). *Comp. Biochem. Physiol.* 73A:571–580.

Abel, E. F. 1973. Zur Öko-Ethologie des amphibisch lebenden Fisches Alticus saliens (Forster) und von Entomacrodus vermiculatus (Val.) (Blennioidea, Salariidae), unter besonderer Berücksichtigung des Fortpflanzungs verhaltens. *Sitzber. Österr, Akad. Wiss., Math.-Naturw. K1., Abt. I,* 181:137–153.

Adiyodi, R. G. 1988. Reproduction and development. In W. W. Burggren and B. R. McMahon, eds., *Biology of the Land Crabs,* pp. 139–185. Cambridge: Cambridge University Press.

Ahlberg, P. E. 1991. Tetrapod or near-tetrapod fossils from the Upper Devonian of Scotland. *Nature* 354:298–301.

Aldridge, D. W. 1983. Physiological ecology of freshwater prosobranchs. In W. D. Russell-Hunter, ed., *The Mollusca,* vol. 6: Ecology, pp. 329–358. Orlando, Fl.: Academic Press.

Alford, R. A. and R. N. Harris. 1988. Effects of larval growth history on anuran metamorphosis. *Amer. Naturalist* 131:91–106.

Ali, M. A. and M. A. Klyne. 1985. *Vision in Vertebrates.* New York: Plenum Press.

Almond, J. R., ed. 1985. *The Silurian-Devonian fossil record of the Myriapod.* Evolution and Environment in the Late Silurian and early Devonian. Phil. Trans. Royal Soc. London.

Amati, B. B., M. Goldschmidt-Clermont, C. J. A. Wallace et al. 1988. cDNA and deduced amino acid sequences of cytochrome c from Chlamydomonas reinhardtii: Unexpected functional and phylogenetic implications. *J. Mol. Evol.* 28:151–160.

al-Ameri, T. K. 1984. Microstructures of possible early land plants from Tripolitania, North Africa. *Rev. Palaeobot. Palynol.* 40:375–386.

Andrews, E. B. 1987. Excretory systems of molluscs. In E. R. Trueman and M. R. Clarke, eds., *The Mollusca,* vol. 11: Form and Function, pp. 381–448. Orlando, Fl.: Academic Press.

Ashley, M. A., S. M. Reilly, and G. V. Lauder. 1991. Ontogenetic scaling of hindlimb muscles across metamorphosis in the tiger salamander, Ambystoma tigrinum. *Copeia* 1991 (3): 767–776.

Atema, J., R. R. Fay, A. N. Popper et al., eds. 1987. *Sensory Biology of Aquatic Animals.* New York: Springer-Verlag.

Atkinson, D. E. 1992. Functional roles of urea synthesis in vertebrates. *Physiol. Zool.* 65:243–267.

Atkinson, R. J. A. and A. C. Taylor. 1988. Physiological ecology of burrowing decapods. *Symp. Zool. Soc. London* 59:201–226.

Audesirk, T. and G. Audesirk. 1985. Behavior of gastropod molluscs. In A. O. D. Willows, ed., *The Mollusca*, vol. 8: Neurobiology and Behavior, Part 1, pp. 1–94. Orlando, Fl.: Academic Press.

Baird, D. and R. Carroll. 1967. Romeriscus, the oldest known reptile. *Science* 157:56–59.

Banks, H. P. 1975a. The oldest vascular land plants: A note of caution. *Rev. Palaeobot. Palynol.* 20:13–25.

—— 1975b. Early vascular land plants: Proof and conjecture. *BioScience* 25:730–737.

Barnhart, M. C. 1986a. Respiratory gas tensions and gas exchange in active and dormant land snails, Otala lactea. *Physiol. Zool.* 59:733–745.

—— 1986b. Control of acid-base status in active and dormant land snails, Otala lactea (Pulmonata, Helicidae). *J. Comp. Physiol., B* 156:347–354.

—— 1989. Respiratory acidosis and metabolic depression in dormant invertebrates. In A. Malan and B. Canguilhem, eds., *Living in the Cold, II*, pp. 321–331. London: J. Libbey.

—— 1992. Acid-base regulation in pulmonate molluscs. *J. Exp. Zool.* 263:120–126.

Barnhart, M. C. and B. R. McMahon. 1987. Discontinuous carbon dioxide release and metabolic depression in dormant land snails. *J. Exp. Biol.* 128:123–138.

—— 1988. Depression of aerobic metabolism and intracellular pH by hypercapnia in land snails, Otala lactea. *J. Exp. Biol.* 138:289–299.

Beaumont, E. H. 1977. Cranial morphology of the Loxomatidae (Amphibia, Labyrinthodontia). *Phil. Trans. Royal Soc. London* B280:29–101.

Beerbower, J. R. 1985. Early development of continental ecosystems. In B. Tiffney, ed., *Geological Factors and the Evolution of Plants*, pp. 47–91. New Haven, Conn.: Yale University Press.

Behrensmeyer, A. K., J. D. Damuth, W. A. DiMichelle et al. 1992. *Terrestrial Ecosystems Through Time.* Chicago: University of Chicago Press.

Bemis, W. E., W. W. Burggren, and N. E. Kemp, eds. 1987. *The Biology and Evolution of Lungfishes.* New York: A. R. Liss.

Benton, M. J., ed. 1988. *The Phylogeny and Classification of the Tetrapod*, vol. 1: Amphibians, Reptiles, Birds. Oxford: Oxford University Press.

Berman, D. S., S. S. Sumida, and R. E. Lombard. 1992. Reinterpretations of the temporal and occipital regions in Diadectes and the relationships of Diadectomorpha. *J. Paleontology* 66:481–499.

Bicudo, J. E. P. W. and E. R. Weibel. 1987. Respiratory exchanges in air breathers versus water breathers: Design contrasts. In P. Dejours, L. Bolis, C. R. Taylor, and E. R. Weibel, eds., *Comparative Physiology*, pp. 15–26. Padova, Italy: Liviana Press.

Bierhorst, D. W. 1971. *Morphology of Vascular Plants*. New York: Macmillan.

Bliss, D. E. 1968. Transitions from water to land in decapod crustaceans. *Am. Zoologist* 8:355–392.

—— 1979. From sea to tree: Saga of a land crab. *Am. Zoologist* 19:385–410.

—— ed. 1982–85. *The Biology of Crustacea*, vols. 1–10. New York: Academic Press.

Bold, H. C. and M. J. Wynne. 1984. *Introduction to the Algae*. Princeton, New Jersey: Prentice Hall.

Bolt, J. R., R. M. McKay, B. J. Witske et al. 1988. A new Lower Carboniferous tetrapod locality in Iowa. *Nature* 333:768–770.

Bond, C. E. 1979. *Biology of Fishes*. Philadelphia: W. B. Saunders.

Boucot, A. J. and C. Janis. 1983. Environment of the early Paleozoic vertebrates. *Palaeogeog. Palaeoclimatol. Palaeoecol.* 41:251–287.

Boudreaux, H. B. 1987. *Arthropod Phylogeny with Special Reference to Insects*. Malabar, Fl.: R. E. Krieger.

Boutilier, R. G., D. F. Stiffler, and D. P. Toews. 1992. Exchange of respiratory gases, ions and water in amphibious and aquatic amphibians. In M. E. Feder and W. W. Burggren, eds., *Environmental Physiology of the Amphibians*, pp. 81–124. Chicago: University of Chicago Press.

Bower, F. O. 1908. *The Origin of a Land Flora*. London: Macmillan.

Bradford, D. F. 1984. Physiological features of embryonic development in terrestrially breeding plethodontid salamanders. In R. S. Seymour, ed., *Respiration and Metabolism of Embryonic Vertebrates*, pp. 87–98. Dordrecht, Holland: W. Junk.

—— 1990. Incubation time and rate of embryonic development in amphibians: The influence of ovum size, temperature and reproductive mode. *Physiol. Zool.* 63:1157–1180.

Bradford, D. F. and R. S. Seymour. 1985. Energy conservation during the delayed hatching period in the frog Pseudophryne bibroni. *Physiol. Zool.* 58:491–496.

—— 1988a. Influence of water potential on growth and survival of the embryo, and gas conductance of the egg, in a terrestrial breeding frog, Pseudophryne bibroni. *Physiol. Zool.* 61:470–474.

—— 1988b. Influence of environmental PO2 on embryonic oxygen consumption, rate of development, and hatching in the frog Pseudophryne bibroni. *Physiol. Zool.* 61:475–482.

Brainerd, E. L., K. F. Liem, and C. T. Samper. 1989. Air ventilation by recoil aspiration in polypterid fishes. *Science* 246:1593–1595.

Bramble, D. M., K. R. Gordon, G. E. J. Goslow et al. 1989. Vertebrate functional morphology: A tribute to Milton Hildebrand. *Amer. Zoologist* 29:187–362.

Bray, A. A. 1985. The evolution of the terrestrial vertebrates: Environmental and physiological considerations. *Phil. Trans. Roy. Soc. London* B309:289–322.

Breder, C. M. and D. E. Rosen. 1966. *Modes of Reproduction in Fishes*. New York: Amer. Mus. Nat. Hist. Press.

Brekke, D. R., S. D. Hillyard, and R. M. Winokur. 1991. Behavior associated with the water absorption response by the toad, Bufo punctatus. *Copeia* 1991 (2): 393–401.

Bremer, K. C. 1985. Summary of green plant phylogeny and classification. *Cladistics* 1:369–385.

Bremer, K. C., C. J. Humphries, B. D. Mishler et al. 1987. On cladistic relationships in green plants. *Taxon* 36:339–349.

Bridges, C. R. 1988. Respiratory adaptations in intertidal fish. *Amer. Zoologist* 28:79–96.

Briggs, D. E. G. 1978. A new trilobite-like arthropod from the Lower Cambrian, Kinzers Formation, Pennsylvania. *J. Paleontol.* 52:133–140.

Briggs, D. E. G. and E. N. K. Clarkson. 1989. Environmental controls on the taphonomy and distribution of Carboniferous malacostracan crustaceans. *Trans. Roy. Soc. Edinburgh, Earth Sci.* 80:293–301.

Briggs, D. E. G., E. N. K. Clarkson, and R. J. Aldridge. 1983. The conodont animal. *Lethaia* 16:1–14.

Briggs, D. E. G. and R. A. Fortey. 1989. The early radiation and relationships of major Arthropod groups. *Science* 246:241–243.

Briggs, D. E. G. and W. D. I. Rolfe. 1983. A giant arthropod trackway from the Lower Mississippian of Pennsylvania. *J. Paleontol.* 57:377–390.

Briggs, D. E. G., W. D. I. Rolfe, and J. Brannan. 1979. A giant myriapod trail from the Namurian of Arran, Scotland. *Palaeontology* 22:273–291.

Brillet, C. 1976. Structure du terrier, reproduction et comportement des jeunes chez le poisson amphibie, Periophthalmus sobrinus Eggert. *Rev. Ecol. (Terre Vie)* 30:465–483.

—— 1986. Notes sur le comportement du poisson amphibie Lophalticus kirkii Gunther (Pisces-Salariidae): Comparaison avec Periophthalmus sobrinus Eggert. *Rev. Ecol. (Terre Vie)* 41:361–376.

Brooks, S. P. J. and K. B. Storey. 1990. Glycolytic enzyme binding and metabolic control in estivation and anoxia in the land snail Otala lactea. *J. Exp. Biol.* 151:193–204.

Brown, A. C. and A. McLachlan, eds. 1990. *Ecology of Sandy Shores*. New York: Elsevier.

Brown, C. R., M. S. Gordon, and H. G. Chin. 1991. Field and laboratory observations on microhabitat selection in the amphibious Red Sea rockskipper fish, Alticus kirki (Family Ble nniidae). *Mar. Behav. Physiol.* 19:1–13.

Brown, C. R., M. S. Gordon, and K. L. M. Martin. 1992. Aerial and aquatic oxygen uptake in the amphibious Red Sea rockskipper fish, Alticus kirki (Family Blenniidae). *Copeia* 1992 (4): 1007–1013.

Bullock, T. H. and W. Heiligenberg, eds. 1986. *Electroreception.* Somerset, New Jersey: J. Wiley & Sons.

Burggren, W. W. 1992. Respiration and circulation in land crabs: Novel variations on the marine design. *Amer. Zoologist* 32:417–427.

Burggren, W. W. and W. E. Bemis. 1990. Studying physiological evolution: Paradigms and pitfalls. In M. H. Nitecki, ed., *Evolutionary Innovations,* pp. 191–228. Chicago: University of Chicago Press.

Burggren, W. W., R. L. J. Infantino, and D. S. Townsend. 1990. Developmental changes in cardiac and metabolic physiology of the direct-developing tropical frog Eleutherodactylus coqui. *J. Exp. Biol.* 152:129–147.

Burggren, W. W. and J. J. Just. 1992. Developmental changes in physiological system. In M. E. Feder and W. W. Burggren, eds., *Environmental Physiology of the Amphibians,* pp. 467–530. Chicago: University of Chicago Press.

Burggren, W. W. and B. R. McMahon. 1988. Circulation. In W. W. Burggren and B. R. McMahon, eds., *Biology of the Land Crabs,* pp. 298–332. Cambridge: Cambridge University Press.

—— eds. 1988. *Biology of the Land Crabs.* Cambridge: Cambridge University Press.

Burggren, W. W. and A. W. Pinder. 1991. Ontogeny of cardiovascular and respiratory physiology in lower vertebrates. *Annu. Rev. Physiol.* 53:107–135.

Burnett, L. E. 1988. Physiological responses to air exposure: Acid-base balance and the role of branchial water stores. *Amer. Zoologist* 28:125–135.

Burton, R. F. 1983. Ionic regulation and water balance. In A. S. M. Saleuddin and K. M. Wilbur, eds., *The Mollusca,* vol. 5: Physiology, Part 2, pp. 291–352. New York: Academic Press.

Butterfield, N. S. 1990. A reassessment of the enigmatic Burgess Shale fossil Wiwaxia corrugata (Matthew) and its relationship to the polychaete Canadia spinosa Walcott. *Paleobiology* 16:287–303.

Cain, A. J. 1983. Ecology and ecogenetics of terrestrial molluscan populations. In W. D. Russell-Hunter, ed., *The Mollusca,* vol. 6: Ecology, pp. 597–647. Orlando, Fl.: Academic Press.

Calow, P. 1983. Life-cycle patterns and evolution. In W. D. Russell-Hunter, ed., *The Mollusca,* vol. 6: Ecology, pp. 649–678. Orlando, Fl.: Academic Press.

Cameron, J. N. 1989. *The Respiratory Physiology of Animals.* New York: Oxford University Press.

274   References

Campbell, K. S. W. and M. W. Bell. 1977. A primitive amphibian from the Late Devonian of New South Wales. *Alcheringa* 1:369–381.

Carefoot, T. H. 1987. Gastropoda. In T. J. Pandian and F. J. Vernberg, eds., *Animal Energetics*, vol. 2: Bivalivia Through Reptilia, pp. 89–172. San Diego: Academic Press.

Carpenter, F. M. 1977. Geological History and Evolution of the Insects. *Proc. 15th Int. Cong. Entomology, Washington, D.C.* 1976:63–70.

Carrier, D. R. 1987. The evolution of locomotor stamina in tetrapods: Circumventing a mechanical constraint. *Paleobiology* 13:326–341.

Carroll, R., E. S. Belt, D. L. Dinely et al. 1972. *Vertebrate paleontology of eastern Canada.* International Geological Congress. Field Excursion A59. Guide Book XXIV.

Carroll, R. L. 1992. The primary radiation of terrestrial vertebrates. *Ann. Rev. Earth Planetary Sci.* 20:45–84.

Cave, L. D. and A. M. Simonetta. 1976. Notes on the morphology and taxonomic position of Aysheaia (Onychophora?) and Skania (indeterminate phylum). *Monitore Zool. Ital.* 9:67–81.

Chaloner, W. G. 1985. *Phil. Trans. Roy. Soc. London* B309:192.

Chaloner, W. G. and J. D. Lawson, eds. 1985. Evolution and environment in the late Silurian and early Devonian. *Phil. Trans. Royal Soc. London* B309:1–342.

Chaloner, W. G. and Sheerin. 1979. Devonian macrofloras. In M. R. House, C. T. Scrutton, and M. G. Bassett, eds., *The Devonian System*, Spec. Pap. Palaeont. No. 23, pp. 145–101. Palaeont. Soc. London.

Chapman, R. L. and M. A. Buchheim. 1992. Green algae and the evolution of land plants: Inferences from nuclear-encoded rRNA gene sequences. *Biosystems* 28:127–137.

Chapman, V. J., ed. 1977. *Wet Coastal Ecosystems.* New York: Elsevier.

Chelazzi, G. and M. Vannini, eds. 1988. *Behavioral Adaptation to Intertidal Life.* New York: Plenum Press.

Cicerone, R. J. 1987. Changes in stratospheric ozone. *Science* 237:35–42.

Clarac, F., F. Libersat, H. J. Pfluger et al. 1987. Motor pattern analysis in the shore crab (Carcinus maenas) walking freely in water and on land. *J. Exp. Biol* 133:395–414.

Cleland, J. D. and R. F. McMahon. 1990. Upper thermal limit of nine intertidal gastropod species from a Hong Kong rocky shore in relation to vertical distribution and desiccation associated with evaporative cooling. In B. Morton, ed., *The Marine Flora and Fauna of Hong Kong and Southern China, II*, vol. 3, pp. 1141–1152. Hong Kong: Hong Kong University Press.

Cloud, P. E. and M. F. Glaessner. 1982. The Ediacaran Period and System: Metazoa inherit the earth. *Science* 218:783–792.

Coates, M. I. and J. A. Clack. 1990. Polydactyly in the earliest known tetrapod limbs. *Nature* 346:66–69.

——— 1991. Fish-like gills and breathing in the earliest known tetrapod. *Nature* 352:234–236.

Conway Morris, S. 1982. *Atlas of Burgess Shale.* Palaeontol. Assoc. London.

——— 1987. The search for the Precambrian-Cambrian boundary. *Amer. Scientist* 75:157–168.

——— 1989. Burgess Shale faunas and the Cambrian explosion. *Science* 246:339–346.

——— 1990. Late Precambrian and Cambrian soft-bodied faunas. *Ann. Rev. Earth Planet. Sci.* 18:101–122.

Conway Morris, S. and R. A. Robison. 1988. *More soft-bodied animals and algae from the Middle Cambrian of Utah and British Columbia.* University Kansas Paleontol. Contrib. Paper 122.

Conway Morris, S. and H. B. Whittington. 1979. The animals of the Burgess Shale. *Sci. American* 240:122–133.

Coombs, S., P. Gorner, and H. Munz, eds. 1989. *The Mechanosensory Lateral Line: Neurobiology and Evolution.* New York: Springer-Verlag.

Costanzo, J. P., M. F. Wright, and R. E. J. Lee. 1992. Freeze tolerance as an overwintering adaptation in Cope's grey treefrog (Hyla chrysoscelis). *Copeia* 1992 (2): 565–569.

Cotter, E. 1978. The evolution of fluvial style, with special reference to the central Appalachian Paleozoic. In A. D. Miall, ed., *Fluvial Sedimentology,* pp. 361–384. Calgary: University of Calgary, Canadian Soc. Petroleum Geol.

Crandall-Stotler, B. 1980. Morphogenetic designs and a theory of bryophyte origins and divergence. *BioScience* 30:580–585.

——— 1986. Morphogenesis, developmental anatomy and bryophyte phylogenetics: Contra-indications of monophyly. *J. Bryol.* 14:1–23.

Crowson, R. A. 1970. *Classification and Biology.* London: Heinemann Educational Books.

——— 1985. Comments on Insecta of the Rhynie Chert. *Entomologica Generalis* 11:97–98.

Dalziel, I. W. D. 1992a. Antarctica: A tale of two supercontinents. *Ann. Rev. Earth Planetary Sci.* 20:501–521.

——— 1992b. On the organization of the American plates, in the Neoproterozoic and the breakout of Laurentia. *Geological Society Today: Geol. Soc. America* 2:237–241.

Daniel, T. L. and P. W. Webb. 1987. Physical determinants of locomotion. In P. Dejours, L. Bolis, C. R. Taylor, and E. R. Weibel, eds., *Comparative Physiology,* pp. 343–369. Padova, Italy: Liviana Press.

Davenport, J. and M. D. J. Sayer. 1986. Ammonia and urea excretion in the amphibious teleost Blennius pholis (L.) in sea water and air. *Comp. Biochem. Physiol.* 84A:189–194.

de Fur, P. L. 1988. Systemic respiratory adaptations to air exposure in intertidal decapod crustaceans. *Amer. Zoologist* 28:115–124.

De Jesus, M. A., F. Tabatabai, and D. J. Chapman. 1989. Taxonomic distribution of copper-zinc superoxide dismutase in green algae and its phylogenetic importance. *J. Phycol.* 25:767–772.

Dejours, P. 1987a. Water and air physical characteristics and their physiological consequences. In P. Dejours, L. Bolis, C. R. Taylor, and E. R. Weibel, eds., *Comparative Physiology*, pp. 3–11. Padova, Italy: Liviana Press.

—— 1987b. Respiratory regulations in water and air breathers: Physiological contrasts. In P. Dejours, L. Bolis, C. R. Taylor, and E. R. Weibel, eds., *Comparative Physiology*, pp. 71–81. Padova, Italy: Liviana Press.

—— 1988. *Respiration in Water and Air: Adaptations—Regulation—Evolution.* Amsterdam: Elsevier.

Dejours, P., L. Bolis, C. R. Taylor et al., eds. 1987. *Comparative Physiology.* Berlin: Springer-Verlag.

del Pino, E. M. 1989. Modifications of oogenesis and development in marsupial frogs. *Development* 107:169–187.

Delwiche, C. F., L. E. Graham, and N. Thomson. 1989. Lignin-like compounds and sporopollenin in Coleochaete, an algal model for land plant ancestry. *Science* 245:399–401.

Denison, R. H. 1956. A review of the habitat of the earliest vertebrates. *Fieldiana: Geology* 11:359–457.

Dent, J. N. 1988. Hormonal interaction in amphibian metamorphosis. *Amer. Zoologist* 28:297–308.

Devereux, R., A. R. Loeblich III, and G. E. Fox. 1990. Higher plant origins and the phylogeny of green algae. *J. Mol. Evol.* 31:18–24.

Doran, J. B. 1980. A new species of Psilophyton from the lower Devonian of Northern New Brunswick, Canada. *Canad. J. Bot.* 58:2241–2262.

Douglas, R. H. and M. B. A. Djamgoz. 1990. *The Visual System of Fish.* London: Chapman and Hall.

Duckett, J. G. 1986. Ultrastructure in bryophyte systematics and evolution: An evaluation. *J. Bryol.* 14:25–42.

Duckett, J. G. and K. S. Renzaglia. 1988. Cell and molecular biology of bryophytes: Ultimate limits to the resolution of phylogenetic problems. *J. Bot. Linn.* 98:225–246.

Dudley, R., V. A. King, and R. J. Wassersug. 1991. The implications of shape and metamorphosis for drag forces on a generalized pond tadpole (Rana catesbeiana). *Copeia* 1991 (1): 252–257.

Duellman, W. E. and L. Trueb. 1986. *Biology of Amphibians.* New York: McGraw-Hill.

Dunham, D. W. and S. L. Gilchrist. 1988. Behavior. In W. W. Burggren and B. R. McMahon, eds., *Biology of the Land Crabs*, pp. 97–138. Cambridge: Cambridge University Press.

Dupré, R. K. and J. W. Petranka. 1985. Ontogeny of temperature selection in larval amphibians. *Copeia* 1985 (2): 462–467.

Dzik, J. and G. Krumbiegel. 1989. The oldest 'Onychophoran' Xenusion: A link connecting phyla? *Lethaia* 22:169–181.

Echteler, S. M. and W. M. Saidel. 1981. Forebrain connections in the goldfish support telencephalic homologies with land vertebrates. *Science* 212:683–685.

Edney, E. B. 1968. Transition from water to land in isopod crustaceans. *Amer. Zoologist* 8:309–326.

Edwards, D. 1970. Fertile Rhyniophytina from the lower Devonian of Britain. *Palaeontology* 13:451–461.

—— 1980. Early land floras. In A. L. Panchen, ed., *The Terrestrial Environment and the Origin of Land Vertebrates*, pp. 55–85. London: Academic Press.

—— 1982. Fragmentary non-vascular plant microfossils from the late Silurian of Wales. *Bot. J. Linn. Soc.* 84:223–256.

Edwards, D., M. G. Bassett, and E. C. W. Rogerson. 1979. The earliest vascular plants: Continuing the search for proof. *Lethaia* 12:313–324.

Edwards, D. and E. C. W. Davies. 1976. Oldest recorded in situ tracheids. *Nature* 263:494–495.

Edwards, D., K. L. Davies, and L. Axe. 1992. A vascular conducting strand in the early land plant Cooksonia. *Nature* 357:683–685.

Edwards, D. and D. S. Edwards. 1986. A reconsideration of the Rhyniophytina Banks. In R. A. Spicer and B. A. Thomas, eds., *Systematic and Taxonomic Approaches in Palaeobotany*, pp. 199–220. Oxford: Clarendon Press.

Edwards, D., U. Fanning, and J. B. Richardson. 1986. Stomata and sterome in early land plants. *Nature* 323:438–440.

Edwards, D. and J. Feehan. 1980. Records of Cooksonia type sporangia from late Wenlock Strata in Ireland. *Nature* 287:41–42.

Edwards, D., J. Feehan, and P. G. Smith. 1983. A late Wenlock flora from Co. Tipperary, Ireland. *Bot. J. Linn. Soc.* 86:19–36.

Edwards, D. and V. Rose. 1984. Cuticles of Nematothallus: A further enigma. *Bot. J. Linn. Soc* 88:35–54.

Edwards, D. S. 1986. Aglaophyton major, a non-vascular plant from the Devonian Rhynie Chert. *Bot. J. Linn. Soc.* 93:173–204.

Edwards, D. S. and A. G. Lyon. 1983. Algae from the Rhynie Chert. *Bot. J. Linn. Soc.* 86:37–55.

Edwards, J. L. 1989. Two perspectives on the evolution of the tetrapod limb. *Amer. Zoologist* 29:235–254.

Elliot, G. F. 1989. The evidence of reproductive mechanisms in fossil dasyclads (algae: Chlorophyta). *Biol. Revs.* 64:269–275.

Erickson, M. and G. E. Miksche. 1974. On the occurrence of lignin or polyphenols in some mosses and liverworts. *Phytochem.* 13:2295–2299.

Evans, D. H. 1987. Acid-base balance and nitrogenous waste excretion in fishes: The aquatic to amphibious transition. In P. Dejours, L. Bolis, C. R. Taylor, and E. R. Weibel, eds., *Comparative Physiology*, pp. 121–129. Padova, Italy: Liviana Press.

Ewert, J.-P. and M. A. Arbib, eds. 1987. *Visuomotor Coordination: Amphibians, Comparisons, Models and Robots*. New York: Plenum Press.

Feder, M. E. and D. T. Booth. 1992. Hypoic boundary layers surrounding skin-breathing aquatic amphibians: Occurrence, consequences and organismal responses. *J. Exp. Biol.* 166:237–251.

Feder, M. E. and W. W. Burggren, eds. 1992. *Environmental Physiology of the Amphibians.* Chicago: University of Chicago Press.

Fedonkin, M. A. 1985. Precambrian metazoans: The problems of preservation, systematics and evolution. *Phil. Trans. Royal Soc. London* B311:27–45.

Field, G. F., G. J. Olson, D. L. Lane et al. 1988. Molecular phylogeny of the animal kingdom. *Science* 239:748–753.

Fisher, D. 1979. Evidence for subaerial activity of Euproops danae (Merostomata: Xiphosuridae). In M. H. Nitecki, ed., *Mazon Creek Fossils,* pp. 379–447. New York: Academic Press.

Fisher, W. A. 1978. The habitats of the early vertebrates: Trace and body fossil evidence from the Harding Formation (Middle Ordovician), Colorado. *Mountain Geol.* 15:1–26.

Flanigan, J. E., P. C. Withers, and M. Guppy. 1991. In vitro metabolic depression of tissues from the aestivating frog Neobatrachus pelobatoides. *J. Exp. Biol.* 161:273–283.

Foreman, R. E., A. Gorbman, J. M. Dodd et al., eds. 1989. *Evolutionary Biology of Primitive Fishes.* New York: Plenum Press.

Frakes, L. A. 1979. *Climates Through Geologic Time.* Amsterdam: Elsevier.

Fretter, V. 1984. Prosobranchs. In A. S. Tompa, N. H. Verdonk, and J. A. M. van den Biggelaar, eds., *The Mollusca,* vol. 7: Reproduction, pp. 1–45. Orlando, Fl.: Academic Press.

Fricke, H. and K. Hissmann. 1992. Locomotion, fin coordination and body form of the living coelacanth Latimeria chalumnae. *Env. Biol. Fishes* 34:329–356.

Fricke, H., O. Reinicke, H. Hofer et al. 1987. Locomotion of the coelacanth Latimeria chalumnae in its natural environment. *Nature* 329:331–333.

Fricke, H., J. Schauer, K. Hissmann et al. 1991. Coelacanth Latimeria chalumnae aggregates in caves: First observations of their resting habitat and social behavior. *Env. Biol. Fishes* 30:281–285.

Friend, J. A. and A. M. M. Richardson. 1986. Biology of terrestrial amphipods. *Ann. Rev. Entomol.* 31:25–48.

Fritsch, F. E. 1916. The algal ancestry of the higher plants. *New Phytol.* 15:233–250.

—— 1921. Thalassiophyta and the algal ancestry of the higher plants. *New Phytol.* 20:165–178.

—— 1945. Studies on the comparative morphology of the algae IV. *Ann. Bot. N. S.* 9:1–29.

Fritzsch, B. 1987. Inner ear of the coelacanth fish Latimeria has tetrapod affinities. *Nature* 327:153–154.

—— 1990. The evolution of metamorphosis in amphibians. *J. Neurobiol.* 21:1011–1021.

Fritzsch, B., M. Ryan, W. Wilczynski et al. 1988. *The Evolution of the Amphibian Auditory System.* Somerset, New Jersey: J. Wiley & Sons.

Frolich, L. M. and A. A. Biewener. 1992. Kinematic and electromyographic analysis of the functional role of the body axis during terrestrial and aquatic locomotion in the salamander Ambystoma tigrinum. *J. Exp. Biol.* 162:107–130.

Full, R. J., B. D. Anderson, C. M. Finnerty et al. 1988. Exercising with and without lungs. I. The effects of metabolic cost, maximal oxygen transport and body size on terrestrial locomotion in salamander species. *J. Exp. Biol.* 138:471–485.

Full, R. J. and R. B. Weinstein. 1992. Integrating the physiology, mechanics and behavior of rapid running ghost crabs: Slow and steady doesn't always win the race. *Amer. Zoologist* 32:382–395.

Galton, V. A. 1988. The role of thyroid hormone in amphibian development. *Amer. Zoologist* 28:309–318.

Gans, C. 1989. Stages in the origin of vertebrates: Analysis by means of scenarios. *Biol. Rev.* 64:221–268.

Gans, C. and G. de Gueldre. 1992. Striated muscle: Physiology and functional morphology. In M. E. Feder and W. W. Burggren, eds., *Environmental Physiology of the Amphibians*, pp. 277–313. Chicago: University of Chicago Press.

Garratt, M. J. 1978. New evidence for a Silurian (Ludlow) age for the earliest Baragwanathia flora. *Alcheringia* 2:217–224.

Garratt, M. J. and R. B. Rickards. 1984. Graptolite biostratigraphy of early land plants from Victoria, Australia. *Proc. Yorks. Geol. Soc.* 44:377–384.

Garratt, M. J., J. D. Tims, J. B. Rickards et al. 1984. The appearance of Baragwanathia (Lycophytina) in the Silurian. *Bot. J. Linn. Soc.* 89:355–358.

Gatten, R. E., Jr., K. Miller, and R. J. Full. 1992. Energetics at rest and during locomotion. In M. E. Feder and W. W. Burggren, eds., *Environmental Physiology of the Amphibians*, pp. 314–377. Chicago: University of Chicago Press.

Geiser, F. and R. S. Seymour. 1989. Influence of temperature and water potential on survival of hatched, terrestrial larvae of the frog Pseudophryne bibronii. *Copeia* 1989 (1): 207–209.

Gensell, P. G. 1977. Morphologic and taxonomic relationships of the Psilotaceae relative to evolutionary lines in early land vascular plants. *Brittonia* 29:14–29.

Gensell, P. G. and H. N. Andrews. 1984. *Plant Life in the Devonian.* New York: Prager Scientific.

—— 1987. The evolution of early land plants. *Amer. Scientist* 75:478–489.

Gensell, P. G., N. G. Johnson, and P. K. Strother. 1990. Early land plant debris (Hooker's "waifs and strays"?). *Palaios* 5:520–547.

Gibson, R. N. 1982. Recent studies on the biology of intertidal fishes. *Oceanogr. Mar. Biol. Annu. Rev.* 20:363–414.

—— 1986. Intertidal teleosts: Life in a fluctuating environment. In T. J. Pitcher, ed., *The Behavior of Teleost Fishes*, pp. 388–408. Baltimore: Johns Hopkins University Press.

Gibson, R. N. 1988. Patterns of movement in intertidal fishes. In G. Chelazzi and M. Vannini, eds., *Behavioral Adaptation to Intertidal Life*, pp. 55–63. New York: Plenum Press.

Gifford, E. M. and A. S. Foster. 1989. *Morphology and Evolution of Vascular Plants*. New York: W. H. Freeman.

Gleeson, T. T. 1991. Patterns of metabolic recovery from exercise in amphibians and reptiles. *J. Exp. Biol.* 160:187–207.

Goldsmith, T. H. 1990. Optimization, constraint, and history in the evolution of eyes. *Quart. Rev. Biol.* 65:281–322.

Gordon, M. S., D. J. Gabaldon, and A. Y. Yip. 1985. Exploratory observations on microhabitat selection within the intertidal zone by the Chinese mudskipper fish Periophthalmus cantonensis. *Marine Biol.* 75:1–7.

Gordon, M. S., W. W. Ng, and A. Y. Yip. 1978. Aspects of the physiology of terrestrial life in amphibious fishes. III. The Chinese mudskipper Periophthalmus cantonensis. *J. Exp. Biol.* 72:57–75.

Gould, S. J. 1989. *Wonderful Life*. New York: Norton.

Gould, S. J. and E. S. Vrba. 1982. Exaptation—a missing term in science of form. *Paleobiology* 8:4–15.

Graffin, G. 1992. A new locality of fossiliferous Harding sandstone: Evidence for fresh water Ordovician vertebrates. *Vert. Paleontol.* 12:1–10.

Graham, A. 1985. Evolution within the Gastropoda: Prosobranchia. In E. R. Trueman and M. R. Clarke, eds., *The Mollusca*, vol. 10: Evolution, pp. 151–186. Orlando, Fl.: Academic Press.

Graham, J. B. 1988. Ecological and evolutionary aspects of integumentary respiration: Body size, diffusion and the Invertebrata. *Amer. Zoologist* 28:1031–1045.

Graham, L. E. 1982. The occurrence and phylogenetic significance of parenchyma in Coleochaete Breb. *Amer. J. Bot.* 69:447–454.

—— 1984. Coleochaete and the origin of land plants. *Amer. J. Bot.* 71:603–608.

—— 1985. The origin of the life cycle of land plants. *Amer. Scientist* 73:178–186.

Graham, L. E. and Y. Kaneko. 1991. Subcellular structures of relevance to the origin of the land plants (embryophytes) from green algae. *Crit. Revs. Plant Sci.* 10:323–342.

Gray, J. 1985a. The microfossil record of early land plants: Advances in understanding of early terrestrialization, 1970–1984. *Phil. Trans. Roy. Soc. London* B309:167–195.

—— 1985b. Ordovician-Silurian land plants: The interdependence of ecology and evolution. *Spec. Papers Palaeontol.* 32:281–295.

—— 1988. Evolution of the freshwater ecosystem: The fossil record. *Palaeogeog. Palaeoclimatol. Palaeoecol.* 62:1–214.

—— 1988. Land plant spores and the Ordovician-Silurian boundary. *Bull. Br. Mus. Nat. Hist. (Geol.)* 43:351–358.

Gray, J. and A. J. Boucot. 1977a. Early Silurian spore tetrads from New York: Earliest New World evidence for vascular plants? *Science* 173:918–921.

—— 1977b. Early vascular land plants: Proof and conjecture. *Lethaia* 10:145–174.

—— 1979. The Devonian land plant Protosalvinia. *Lethaia* 12:57–63.

Gray, J., D. Massa, and A. J. Boucot. 1982. Caradocian land plant microfossils from Libya. *Geology* 10:197–201.

Gray, J. and W. Shear. 1992. Early life on land. *Amer. Scientist* 80:444–456.

Gray, J., J. N. Theron, and A. J. Boucot. 1986. Age of the Cedarberg Formation, South Africa, and early land plant evolution. *Geol. Mag.* 123:445–454.

Greenaway, P. 1985. Calcium balance and moulting in the Crustacea. *Biol. Rev.* 60:425–454.

—— 1988. Ion and water balance. In W. W. Burggren and B. R. McMahon, eds., *Biology of the Land Crabs*, pp. 211–248. Cambridge: Cambridge University Press.

Greenaway, P. and C. Farrelly. 1990. Vasculature of the gas-exchange organs in air-breathing brachyurans. *Physiol. Zool.* 63:117–139.

Greenaway, P. and S. Morris. 1989. Adaptations to a terrestrial existence by the robber crab, Birgus latro L., III. Nitrogenous excretion. *J. Exp. Biol.* 143:333–346.

Greenaway, P., S. Morris, and B. R. McMahon. 1988. Adaptations to a terrestrial existence by the robber crab Birgus latro L., II. In vivo respiratory gas exchange and transport. *J. Exp. Biol* 140:493–509.

Greenaway, P. and T. Nakamura. 1991. Nitrogenous excretion in two terrestrial crabs (Gecarcoidea natalis and Geograpsus grayi). *Physiol. Zool.* 64:767–786.

Greenaway, P., H. H. Taylor, and S. Morris. 1990. Adaptations to a terrestrial existence by the robber crab, Birgus latro L., VI. The role of the excretory system in fluid balance. *J. Exp. Biol.* 152:505–519.

Greenslade, P. J. M. 1988. Reply to R. A. Crowson's "Comments on Insecta of the Rhynie Chert." *Entomologica Generalis* 13:115–117.

Griffith, R. W. 1991. Guppies, toadfish, lungfish, coelacanths and frogs: A scenario for the evolution of urea retention in fishes. *Env. Biol. Fishes* 32:199–218.

Grote, J. R. 1981. The effect of load on locomotion in crayfish. *J. Exp. Biol.* 92:277–288.

Haeckel, E. 1866. *Generelle Morphologie der Organismen.* Berlin: Reimer.

Han, T.-M. and B. Runnegar. 1992. Megascopic eukaryotic algae from the 2.1-billion-year-old Negaunee iron-formation, Michigan. *Science* 257:232–234.

Hanke, W., ed. 1990. *Biology and Physiology of Amphibians.* Stuttgart, Germany: G. Fischer Verlag.

Hanken, J. 1989. Development and evolution in amphibians. *Amer. Scientist* 77:336–343.

Harland, W. B., R. L. Armstrong, A. V. Cox et al. 1990. *A Geologic Time Scale 1989*. Cambridge: Cambridge University Press.

Hart, C. W., Jr. and J. Clark. 1989. *An Interdisciplinary Bibliography of Freshwater Crayfishes (Astacoidea and Parastacoidea) from Aristotle through 1985 and updated through 1987*. Washington, D.C.: Smithsonian Institution Press.

Hartnady, C. J. H. 1991. About turn for supercontinents. *Nature* 352:476–478.

Hartnoll, R. G. 1988. Evolution, systematics and geographical distribution. In W. W. Burggren and B. R. McMahon, eds., *Biology of the Land Crabs*, pp. 6–54. Cambridge: Cambridge University Press.

—— 1988. Growth and molting. In W. W. Burggren and B. R. McMahon, eds., *Biology of the Land Crabs*, pp. 186–210. Cambridge: Cambridge University Press.

Hennig, W. 1981. *Insect Phylogeny*. New York: Academic Press.

Henry, R. P. 1991. Branchial and branchiostegite carbonic anhydrase in decapod crustaceans: The aquatic to terrestrial transition. *J. Exp. Zool.* 259:294–303.

Herak, M., V. Kochansky-Devidé and I. Gusic. 1977. Development of the Dasyclad algae through the ages. In E. Flügel, ed., *Fossil Algae*, pp. 143–153. Berlin: Springer-Verlag.

Herreid, C. F., II and R. J. Full. 1988. Energetics and locomotion. In W. W. Burggren and B. R. MacMahon, eds., *Biology of the Land Crabs*, pp. 334–377. Cambridge: Cambridge University Press.

Hesselbo, S. P. 1992. Agalaspida (Arthropoda) from the Upper Cambrian of Wisconsin. *J. Paleontol.* 66:885–923.

Heyler, D. 1985–86. Sur le Gisement Stephanien de Montceau les Mines. Table Ronde International du CNRS. *Bull. Trimest. Mus. d'Autun* 114–117:1–173.

Higgins, R. P. and H. Thiel, eds. 1988. *Introduction to the Study of Meiofauna*. Washington, D.C.: Smithsonian Institution Press.

Hilbish, T. J. 1981. Latitudinal variation in freezing tolerance of Melampus bidentatus (Say) (Gastropoda: Pulmonata). *J. Exp. Mar. Biol. Ecol.* 52:283–297.

Hildebrand, M., D. M. Bramble, K. F. Liem et al., eds. 1985. *Functional Vertebrate Morphology*. Cambridge, Mass.: Harvard University Press.

Hillerton, J. E. 1984. Arthropoda: Cuticle: Mechanical properties. In J. Bereiter-Hahn, A. G. Matoltsy, and K. S. Richards, eds., *Biology of the Integument*, vol. 1: Invertebrates, pp. 626–637. Berlin: Springer-Verlag.

Hillis, D. M. 1991. The phylogeny of amphibians: Current knowledge and the role of cytogenetics. In D. M. Green and S. K. Sessions, eds., *Amphibian Cytogenetics and Evolution*, pp. 7–31. San Diego: Academic Press.

Hillman, S. S. 1988. Dehydrational effects on brain and cerebrospinal fluid electrolytes in two amphibians. *Physiol. Zool.* 61:254–259.

—— 1991. Cardiac scope in amphibians: Transition to terrestrial life. *Can. J. Zool.* 69:2010–2013.

Hirst, S. 1922. On some arachnid remains from the Old Red Sandstone (Rhynie Chert beds, Aberdeenshire). *Ann. Mag. Nat. Hist.* 12:455–474.

Hirst, S. and B. Maulik. 1926. On some arthropod remains from the Rhynie Chert (Old Red Sandstone). *Geol. Mag.* 63:69–71.

Hobbs, H. H., Jr. 1981. *The Crayfishes of Georgia.* Washington, D.C.: Smithsonian Institution Press.

Hoek, C. v. d., W. T. Stam, and J. L. Olsen. 1988. The emergence of a new chlorophytan system, and Dr. Kornmann's contribution thereto. *Helgoländ. Meeresunters.* 42:339–383.

Holdich, D. M. 1984. The cuticular surface of woodlice: A search for receptors. *Symp. Zool. Soc. London* 53:9–48.

Holland, H. D., B. Lazar, and M. McCaffrey. 1986. Evolution of the atmosphere and oceans. *Nature* 320:27–33.

Holmgren, N. K. 1939. Contributions to the question of the origin of the tetrapod limb. *Acta Zool.* 20:89–124.

Hori, H., B.-L. Lim, and S. Osawa. 1985. Evolution of green plants as deduced from 5S rRNA sequences. *Proc. Nat. Acad. Sci. USA.* 82:820–823.

Hori, H. and S. Osawa. 1986. Evolutionary change in 5S rRNA species. *Biosystems* 19:163–172.

Horn, M. H., M. A. Neighbors, and S. N. Murray. 1986. Herbivore responses to a seasonally fluctuating food supply: Growth potential of two temperate intertidal fishes based on the protein and energy assimilated from their macroalgal diets. *J. Exp. Mar. Biol. Ecol.* 103:217–234.

Horn, M. H. and K. C. Riegle. 1981. Evaporative water loss and intertidal vertical distribution in relation to body size and morphology of stichaeoid fishes from California. *J. Exp. Mar. Biol. Ecol.* 50:273–288.

Horwitz, P. H. J. and A. M. M. Richardson. 1986. An ecological classification of the burrows of Australian freshwater crayfish. *Aust. J. Mar. Freshwat. Res.* 37:237–242.

House, M. R., J. B. Richardson, W. G. Chaloner et al. 1977. A correlation of the Devonian Rocks in the British Isles. *Spec. Rep. Geol. Soc. London* 8:1–110.

Hueber, F. M. 1983. A new species of Baragwanathia from the Sextant Formation (Emsian) Northern Ontario, Canada. *Bot. J. Linn. Soc.* 86:57–79.

Hui, C. A. 1992. Walking of the shore crab Pachygrapsus crassipes in its two natural environments. *J. Exp. Biol.* 165:213–227.

Hutchison, V. H. and R. K. Dupré. 1992. Thermoregulation. In M. E. Feder and W. W. Burggren, eds., *Environmental Physiology of the Amphibians,* pp. 206–249. Chicago: University of Chicago Press.

Hyde, D. A., T. W. Moon, and S. F. Perry. 1987. Physiological consequences of prolonged aerial exposure in the American eel: Blood respiratory and acid-base status. *J. Comp. Physiol.* 157B:635–642.

Hyde, D. A. and S. F. Perry. 1987. Acid-base and ionic regulation in the

American eel (Anguilla rostrata) during and after prolonged aerial exposure: Branchial and renal adjustments. *J. Exp. Biol.* 133:429–447.

Ip, Y. K., S. F. Chew, and R. W. L. Lim. 1990. Ammoniagenesis in the mudskipper, Periophthalmus chrysospilos. *Zool. Sci.* 7:187–194.

Ivanov, S. N., A. A. Efimov, G. A. Smirnov et al. 1975. Fundamental features in the structure and evolution of the Urals. *Am. J. Sci.* 275A:107–130.

Iwata, K. 1988. Nitrogen metabolism in the mudskipper, Periophthalmus cantonensis: Changes in free amino acids and related compounds in various tissues under conditions of ammonia loading, with special reference to its high ammonia tolerance. *Comp. Biochem. Physiol.* 91A:499–508.

Jacobs, D. K. 1990. Selector genes and the Cambrian radiation of the Bilateria. *Proc. Nat. Acad. Sci. USA* 87:4406–4410.

Jarvik, E. 1952. On the fish-like tail in the ichthyostegid stegocephalians. *Meddel. om Grønland* 114:1–90.

—— 1955. The oldest tetrapods and their forerunners. *Sci. Monthly* 80:141–154.

—— 1981. Lungfishes, tetrapods, paleontology and plesimorphy (Review). *Syst. Zool* 30:378–384.

Jefferies, R. P. S. 1986. *The Ancestry of the Vertebrates*. Cambridge: Cambridge University Press.

Johnson, R. E. and P. M. Sheehan. 1985. Late Ordovician Dasyclad algae of the Eastern Great Basin. In D. F. T. Nitecki and M. H. Nitecki, eds., *Paleoalgology*, pp. 79–84. Berlin: Springer-Verlag.

Jones, H. D. 1983. The circulatory systems of gastropods and bivalves. In A. S. M. Saleuddin and K. M. Wilbur, eds., *The Mollusca*, vol. 5: Physiology, Part 2, pp. 189–238. New York: Academic Press.

Jørgensen, C. B. 1992. Growth and reproduction. In M. E. Feder and W. W. Burggren, eds., *Environmental Physiology of the Amphibians*, pp. 439–466. Chicago: University of Chicago Press.

Kantz, T. S., E. C. Theriot, E. A. Zimmer et al. 1990. The Pleurastrophyceae and micromonadophyceae: A cladistic analysis of nuclear rRNA sequence data. *J. Phycol.* 26:711–721.

Kasting, J. F. 1993. Earth's early atmosphere. *Science* 259:920–926.

Kasting, J. F., H. D. Holland, and L. R. Kump. 1992. Atmospheric evolution: The rise of oxygen. In J. W. Schopf, ed., *The Proterozoic Earth*, pp. 159–163. Cambridge: Cambridge University Press.

Keller, E. F. and E. A. Lloyd. 1992. *Keywords in Evolutionary Biology*. Cambridge, Mass.: Harvard University Press.

Kethley, J. B., R. A. Norton, D. B. Bonamo et al. 1989. A terrestrial alicorhagid mite (Acari: Acariformes) from the Devonian of New York. *Micropaleontology* 35:367–373.

Kevan, P. G., W. G. Chaloner, and D. B. O. Saville. 1975. Interrelationships of early terrestrial arthropods and plants. *Palaeontology* 18:391–417.

Kidston, R. and W. H. Lang. 1917. On Old Red Sandstone plants showing structure from the Rhynie chert bed, Aberdeenshire, I. Rhynia Gwynne-Vaughanii. *Trans. Roy. Soc. Edinburgh* 51:761–784.

—— 1921. On Old Red Sandstone plants showing structure from the Rhynie Chert Bed, Aberdeenshire, IV. Restorations of the vascular cryptogams and discussion of their bearing on the general morphology of the Pteridophyta and the origin of the organization of land plants. *Trans. Roy. Soc. Edinburgh* 52:831–854.

Kjellesvig-Waering, E. N. 1986. A restudy of the fossil Scorpionida of the world. *Palaeontographica Americana* 5:1–287.

Knoll, A. H., S. W. F. Grant, and J. W. Tsao. 1986. The early evolution of land plants. In R. A. Gastaldo, ed., *Early Land Plants* (Univ. Tennessee Studies in Geology), pp. 45–63. Memphis: University of Tennessee Press.

Knoll, A. H. and G. W. Rothwell. 1981. Paleobotany: Perspectives in 1980. *Paleobiology* 7:7–35.

Kohn, A. J. 1983. Feeding biology of gastropods. In A. S. Saleuddin and K. M. Wilbur, eds., *The Mollusca*, vol. 5: Physiology, Part 2, pp. 1–63. New York: Academic Press.

Kolattukudy, P. E. 1980. Biopolyester membranes of plants: Cutin and Suberin. *Science* 208:990–1000.

Kovâcs-Endrödy, E. 1986. The earliest known vascular plant, or a possible ancestor of vascular plants in the flora of the lower Silurian Cedarberg Formation, Table Mountain Group, South Africa. *Ann. Geol. Surv. South Afr.* 20:93–118.

Kristensen, N. P. 1989. Insect phylogeny based on morphological evidence. In B. Fernholm, K. Breuner, and H. Jörnwall, eds., *The Hierarchy of Life*, pp. 295–306. Amsterdam: Elsevier.

Kryshtovovich, A. N. 1953. Discovery of a clubmoss-like plant in the Cambrian of Eastern Siberia. *Dokl. Akad. Nauk SSR* 91:1377–1379.

Kukalova, J. 1964. To the morphology of the oldest known dragonfly Erasipteron larischi Purvost 1933. *Vstn. Ustred Ustavu Geologie* 39:463–464.

Kukalova-Peck, J. 1990. The "Uniramia" do not exist: The ground plan of the Pterygota as revealed by Permian Diaphanopteroidea from Russia (Insecta: Palaeodictyopteroidea). *Canad. J. Zool.* 70:236–255.

—— 1991. Fossil history and the evolution of hexapod structures. In I. D. Naumann et al., eds., *Insects of Australia*, vol 1, pp. 141–179. Ithaca: Cornell University Press.

Labandeira, C. and B. S. Beall. 1990. Arthropod terrestrialization. In D. Mikulic and S. J. Culver, eds., *Arthropod Paleobiology No. 3: Short Courses in Paleontology*, pp. 214–232. Knoxville, Tenn.: The Paleontological Society.

Labandeira, C. C., B. S. Beall, and F. M. Hueber. 1988. Early insect diversification: Evidence from a Lower Devonian bristletail from Quebec. *Science* 242:193–196.

Lake, J. A. 1990. Origin of the Metazoa. *Proc. Nat. Acad. Sci. USA* 87:763–766.

Land, M. F. 1987. Vision in air and water. In P. Dejours, L. Bolis, C. R.

Taylor, and E. R. Weibel, eds., *Comparative Physiology*, pp. 289–302. Padova, Italy: Liviana Press.

Land, M. F. and R. D. Fernald. 1992. The evolution of eyes. *Ann. Rev. Neurosci.* 15:1–29.

Lang, W. H. 1937. On the plant remains from the Downtonian of England and Wales. *Phil. Trans. Roy. Soc. Lond.* B227:245–291.

Lang, W. H. and I. C. Cookson. 1935. On a flora, including vascular plants, associated with Monograptus, in rocks of Silurian age, from Victoria Australia. *Phil. Trans. Roy. Soc. Lond.* B224:421–449.

Lauder, G. V. 1990. Functional morphology and systematics: Studying functional patterns in an historical context. *Annu. Rev. Ecol. Syst.* 21:317–340.

Lauder, G. V. and S. M. Reilly. 1990. Metamorphosis of the feeding mechanism in tiger salamanders (Ambystoma tigrinum): The ontogeny of cranial muscle mass. *J. Zool., London* 222:59–74.

Lebedev, O. A. 1984. The first find of a Devonian tetrapod vertebrate in the USSR. (In Russian). *Doklady Akad. Sci. USSR* 278:1470–1473.

Lee, R. E. 1989. *Phycology*. Cambridge: Cambridge University Press.

Leonardi, G. 1982. Descoberta e pegada de um amfibio Devoniano no Parana. *Ciencias Terra* 1982:36–37.

—— 1983. Notopus petri nov. gen, nov. sp.: Une empreinte d'amphibien de Devonian au Parana' (Bresil). *Geobios* 16:233–239.

Levine, J. S. 1980. Surface solar ultraviolet radiation for paleoatmospheric levels of oxygen and ozone. *Origins of Life* 10:313–323.

Levine, J. S., R. E. Boughner, and K. A. Smith. 1980. Ozone, ultraviolet flux and temperature of the paleoatmosphere. *Origins of Life* 10:199–213.

Liem, K. F. 1988. Form and function of lungs: The evolution of air breathing mechanisms. *Amer. Zoologist* 28:739–759.

Little, C. 1983. *The Colonisation of Land: Origins and Adaptations of Terrestrial Animals*. Cambridge: Cambridge University Press.

—— 1990. *The Terrestrial Invasion: An Ecophysiological Approach to the Origins of Land Animals*. Cambridge: Cambridge University Press.

Little, C., M. C. Pilkington, and J. B. Pilkington. 1984. Development of salinity tolerance in the marine pulmonate Amphibola crenata (Gmelin). *J. Exp. Mar. Biol. Ecol.* 74:169–177.

Logan, K. J. and B. A. Thomas. 1985. Distribution of lignin derivatives in plants. *New Phytol.* 99:571–585.

Lombard, R. E. 1991. Experiment and comprehending the evolution of function. *Amer. Zoologist* 31:743–756.

Lombard, R. E. and S. S. Sumida. 1992. Recent progress in understanding early tetrapods. *Amer. Zoologist* 32:609–622.

Macintosh, D. J. 1988. The ecology and physiology of decapods of mangrove swamps. *Symp. Zool. Soc. London* 59:315–341.

Maitland, D. P. and A. Maitland. 1992. Penetration of water into blind-ended capillary tubes and its bearing on the functional design of the lungs of soldier crabs Mictyris longicarpus. *J. Exp. Biol.* 163:333–344.

Malvin, G. M. and N. Heisler. 1988. Blood flow patterns in the salamander, Ambystoma tigrinum, before, during and after metamorphosis. *J. Exp. Biol.* 137:53–74.

Mantel, L. H. 1992. The compleat crab. *Amer. Zoologist* 32:361–540.

Manton, S. M. 1977. *The Arthropoda: Habits, Functional Morphology and Evolution.* Oxford: Clarendon Press.

Marshall, C. and H.-P. Schultze. 1992. Relative importance of molecular, neontological and paleontological data in understanding the biology of the vertebrate invasion of land. *J. Molecular Evol.* 35:93–101.

Martin, A. W. 1983. Excretion. In A. S. M. Saleuddin and K. M. Wilbur, eds., *The Mollusca*, vol. 5: Physiology, Part 2, pp. 353–405. New York: Academic Press.

Martin, K. L. M. 1991. Facultative aerial respiration in an intertidal sculpin, Clinocottus analis (Scorpaeniformes: Cottidae). *Physiol. Zool.* 64:1341–1355.

Martin, K. L. M. and J. R. B. Lighton. 1989. Aerial CO_2 and O_2 exchange during terrestrial activity in an amphibious fish, Alticus kirki (Blenniidae). *Copeia* 1989 (3): 723–727.

Mattox, K. R. and K. D. Stewart. 1984. A classification of the green algae: A concept based on comparative cytology. In D. E. G. Irvine and D. M. Johns, eds., *Systematics of the Green Algae*, pp. 29–72. London: Academic Press.

McKenzie, K. G. 1983. On the origin of Crustacea. *Mem. Australian Mus.* 18:21–43.

McKerrow, W. S. and C. R. Scotese. 1990. Palaeozoic palaeogeography and biogeography. *Geol. Soc. London, Memoir* 12:1–435.

McMahon, B. R. and W. W. Burggren. 1988. Respiration. In W. W. Burggren and B. R. McMahon, eds., *Biology of the Land Crabs*, pp. 249–297. Cambridge: Cambridge University Press.

McMahon, B. R. and P. R. H. Wilkes. 1983. Emergence responses and aerial ventilation in normoxic and hypoxic crayfish, Orconectes rusticus. *Physiol. Zool.* 56:133–141.

McMahon, R. F. 1983. Physiological ecology of freshwater pulmonates. In W. D. Russell-Hunter, ed., *The Mollusca*, vol. 6: Ecology, pp. 359–430. Orlando, Fl.: Academic Press.

—— 1988. Respiratory response to periodic emergence in intertidal molluscs. *Amer. Zool.* 28:97–114.

—— 1990. Thermal tolerance, evaporative water loss, air-water oxygen consumption and zonation of intertidal prosobranchs: A new synthesis. *Hydrobiologia* 193:241–260.

McMahon, R. F. and J. C. Britton. 1985. The relationship between vertical distribution, thermal tolerance, evaporative water loss rate, and behaviour on emergence in six species of mangrove gastropods from Hong Kong. In B. Morton and D. Dudgeon, eds., *The Malacofauna of Hong Kong and Southern China, II*, vol. 2, pp. 563–582. Hong Kong: Hong Kong University Press.

McMahon, R. F. and J. D. Cleland. 1990. Thermal tolerance, evaporative water loss and behaviour during prolonged emergence in the high zoned mangrove gastropod Cerithideaornata: Evidence for atmospheric water uptake. In B. Morton, ed., *The Marine Flora and Fauna of Hong Kong, II*, vol. 3, pp. 1123–1139. Hong Kong: Hong Kong University Press.

McMahon, R. F. and W. D. Russell-Hunter. 1981. The effects of physical variables and acclimation on survival and oxygen consumption in the high littoral salt-marsh snail, Melampus bidentatus Say. *Biol. Bull. (Woods Hole)* 161:246–269.

McMenamin, M. A. S. and D. L. S. McMenamin. 1990. *The Emergence of Animals: The Cambrian Breakthrough.* New York: Columbia University Press.

Melkonian, M. 1982. Structural and evolutionary aspects of the flagellar apparatus in green algae. *Taxon* 31:255–265.

—— 1984. Flagellar structure ultrastructure; in relation to green algae classification. In D. E. G. Irvine John and D. M. John, eds., *Systematics of the Green Algae*, pp. 73–120. London: Academic Press.

Messner, B. 1988. Sind die Insekten primäre oder secondäre Wasser-bewohner? *Deut. Entomol. Z., N. F.* 35:355–360.

Mikulic, D. G., D. E. G. Briggs, and J. Klussendorf. 1985. Silurian soft-bodied biota. *Science* 228:715–717.

Miller, B. T. and J. J. H. Larsen. 1990. Comparative kinematics of terrestrial prey capture in salamanders and newts (Amphibia: Urodela: Salamandridae). *J. Exp. Zool.* 256:135–153.

Milner, A. R., T. R. Smithson, A. C. Milner et al. 1986. The search for early tetrapods. *Modern Geology* 10:1–28.

Milsom, W. K. 1991. Intermittent breathing in vertebrates. *Ann. Rev. Physiol.* 53:87–105.

Mishler, B. D. and S. P. Churchill. 1984. A cladistic approach to the phylogeny of the bryophytes. *Brittonia* 36:406–424.

—— 1985. Transition to a land flora: Phylogenetic relationships of the green algae and bryophytes. *Cladistics* 1:305–328.

Moestrup, P. 1978. On the phylogenetic validity of the flagellar apparatus in green algae and other chlorophyll a + b–containing organisms. *Biosystems* 10:117–144.

Mohr, R. E. 1975. Measured periodicity of the Biwalic (Precambrian) stromatolites and their geophysical significance. In G. D. Rosenberg and S. K. Runcorn, eds., *Growth Rhythms and the History of the Earth's Rotation*, pp. 43–56. London: Wiley Interscience.

Mommsen, T. P. and P. J. Walsh. 1989. Evolution of urea synthesis in vertebrates: The piscine connection. *Science* 243:72–75.

—— 1991. Urea synthesis in fishes: Evolutionary and biochemical perspectives. In P. W. Hochachka and T. P. Mommsen, eds., *Biochemistry and Molecular Biology of Fishes*, vol. 1, pp. 137–163. Amsterdam: Elsevier.

Moore, R. C., ed. 1955. *Treatise on Invertebrate Paleontology*. Part P: Arthropoda 2. Geol. Soc. America and University Kansas Press.

Morel, P. and E. Irving. 1978. Tentative paleocontinental maps for the early Phanerozoic and Paleozoic. *J. Geol.* 86:535–562.

Morris, S., H. H. Taylor, and P. Greenaway. 1991. Adaptations to a terrestrial existence by the robber crab, Birgus latro, VII. The branchial chamber and its role in urine reprocessing. *J. Exp. Biol.* 161:315–331.

Morritt, D. 1988. Osmoregulation in littoral and terrestrial talitroidean amphipods (Crustacea) from Britain. *J. Exp. Mar. Biol. Ecol.* 123:77–94.

Morton, J. E. 1988. The pallial cavity. In E. R. Trueman and M. R. Clarke, eds., *The Mollusca*, vol. 11: Form and Function, pp. 253–286. San Diego: Academic Press.

Müller, K. J. 1983. Crustacea with preserved soft parts from the Upper Cambrian of Sweden. *Lethaea* 16:93–109.

Mundel, P. 1979. The centipedes (Chilopoda) of Mazon Creek. In M. H. Nitecki, ed., *Mazon Creek Fossils*, pp. 361–378. New York: Academic Press.

Munro, A., T. J. Lam, and A. P. Scott. 1990. *Reproductive Seasonality in Teleosts: Environmental Influences*. Boca Raton, Fl.: CRC Press.

Murphy, D. J. 1983. Freezing resistance in intertidal invertebrates. *Ann. Rev. Physiol.* 45:289–299.

Musick, J. A., M. N. Bruton, and E. K. Balon. 1991. The biology of Latimeria chalumnae and evolution of coelacanths. *Env. Biol. Fishes* 32:1–433.

Narins, P. 1990. Seismic communication in anuran amphibians. *BioScience* 40:268–274.

Naylor, E. 1988. Rhythmic behavior of decapod crustaceans. *Symp. Zool. Soc. London* 59:177–199.

Nelson, J. S. 1984. *Fishes of the World*. New York: J. Wiley and Sons.

Nevenzel, J. C., W. Rodegker, J. F. Mead et al. 1966. Lipids of the living coelacanth, Latimeria chalumnae. *Science* 152:1753–1755.

Neville, A. C. 1984. Arthropods: Cuticle: Organization. In J. Bereiter-Hahn, A. G. Matoltsy, and K. S. Richards, eds., *Biology of the Integument*, vol. 1: Invertebrates, pp. 611–625. Berlin: Springer-Verlag.

Niklas, K. 1976a. Morphological and ontogenetic reconstruction of Parka decipiens Fleming and Pachytheca Hooker from the lower Old Red Sandstone, Scotland. *Trans. Roy. Soc. Edinburgh* 69:483–499.

—— 1976b. The chemotaxonomy of Parka decipiens from lower Old Red Sandstone, Scotland (U.K.). *Rev. Palaeobot. Palynol.* 21:205–217.

—— 1976c. Chemotaxonomy of Prototaxites and evidence for possible terrestrial adaptation. *Rev. Palaeobot. Palynol.* 22:1–7.

Niklas, K. 1979. An assessment of chemical features for the classification of plant fossils. *Taxon* 28:505–516.

—— 1980. Paleobiochemical techniques and their applications to paleobotany. *Progr. Phytochem.* 7:143–181.

—— 1981. The chemistry of fossil plants. *BioScience* 31:820–825.

—— 1990. Biomechanics of Psilotum nudum and some early paleozoic vascular sporophytes. *Amer. J. Bot.* 77:590–606.

Niklas, K. and W. G. Chaloner. 1976. Chemotaxonomy of some problematic palaeozoic plants. *Rev. Palaeobot. Palynol.* 22:81–104.

Niklas, K. and T. L. Phillips. 1976. Morphology of Protosalvinia from the Upper Devonian of Ohio and Kentucky. *Amer. J. Bot.* 63:9–29.

Niklas, K. and L. M. Pratt. 1980. Evidence for lignin-like constituents in early Silurian (Llandoverian) plant fossils. *Science* 209:396–397.

Niklas, K. and V. Smocovitis. 1983. Evidence for a conducting strand in early Silurian (Llandoverian) plants: Implications for the evolution of land plants. *Paleobiol.* 9:126–137.

Norris, D. O. and R. E. Jones, eds. 1987. *Hormones and Reproduction in Fishes, Amphibians and Reptiles.* New York: Plenum Press.

O'Kelly, C. J. and G. L. Floyd. 1984. Flagella apparatus, absolute orientations and the phylogeny of the green algae. *Biosystems* 16:227–251.

Olson, E. C. 1984. Origine et evolution des communautes biologiques de vertebres Permo-Carboniferes. *Ann. Paleontol.* 70:41–82.

—— 1985. Nonmarine vertebrates and late Paleozoic climates. In *IXth International Carboniferous Congress, Comptes Rendus,* vol. 5, pp. 403–414. Carbondale, Illinois: University of Illinois Press.

Panchen, A. L., ed. 1980. *The Terrestrial Environment and the Origin of Land Vertebrates.* vol. 15, Systematics Assoc. Spec. London: Academic Press.

Panchen, A. L. and T. R. Smithson. 1987. Character diagnosis, fossils and the origin of tetrapods. *Biol. Rev.* 62:341–438.

Peer, Y. v. d., R. de Baere, J. Cauwenberghs et al. 1990. Evolution of green plants and their relationship with other photosynthetic eukaryotes as deduced from 5S ribosomal RNA sequences. *Pl. System. Evol.* 170:85–96.

Phillips, T. L., K. J. Nicklas, and H. N. Andrews. 1972. Morphology and vertical distribution of Protosalvina (Foerstia) from the New Albany Slate (Upper Devonian). *Rev. Palaeobot. Palynol.* 14:171–196.

Pickett-Heaps, J. D. 1975. *Green Algae.* Sunderland, Mass: Sinauer.

Pickett-Heaps, J. D. and H. J. Marchant. 1972. The phylogeny of the green algae: A new proposal. *Cytobios* 6:255–264.

Pilkington, J. B., C. Little, and P. E. Stirling. 1984. A respiratory current in the mantle cavity of Amphibola crenata (Mollusca, Pulmonata). *J. Roy. Soc. New Zealand* 14:327–334.

Pinder, A. W., K. B. Storey, and G. R. Ultsch. 1992. Estivation and hibernation. In M. E. Feder and W. W. Burggren, eds., *Environmental Physiology of the Amphibians,* pp. 250–274. Chicago: University of Chicago Press.

Piper, J. D. A. 1977. Geological and geophysical evidence relating to continental growth and dynamics and the hydrosphere in Precambrian times: A review and analysis. In P. Brosche and J. Sunderman, eds., *Tidal Friction and the Earth's Rotation*, pp. 197–241. Berlin: Springer-Verlag.

—— 1983. Dynamics of the continental crust in Proterozoic times. In L. G. Medaris, Jr., C. W. Byers, D. M. Mickelson, and W. C. Shanks, eds., *Proterozoic Geology*, Geol. Soc. America Memoir 161, pp. 11–34.

Pirozynski, K. A. and D. W. Malloch. 1975. The origin of land plants: A matter of mycotrophysm. *Biosystems* 6:153–164.

Poinar, G. O., Jr. 1992. *Life in Amber*. Stanford, Calif.: Stanford University Press.

Potts, G. W. and R. T. Wootton, eds. 1984. *Fish Reproduction: Strategies and Tactics*. London: Academic Press.

Pough, F. H., J. B. Heiser, and W. N. McFarland. 1989. *Vertebrate Life*. New York: Macmillan.

Pough, F. H., W. E. Magnusson, M. J. Ryan et al. 1992. Behavioral energetics. In M. E. Feder and W. W. Burggren, eds., *Environmental Physiology of the Amphibians*, pp. 395–436. Chicago: University of Chicago Press.

Powers, L. W. and D. E. Bliss. 1983. Terrestrial adaptations. In F. J. Vernberg and W. B. Vernberg, eds., *The Biology of Crustacea*, vol. 8: Environmental Adaptations, pp. 271–333. New York: Academic Press.

Pratt, L. M., T. L. Phillips, and J. M. Dennison. 1978. Evidence of nonvascular land plants from the early Silurian (Llandoverian) of Virginia, USA. *Rev. Palaeobot. Palynol.* 25:121–149.

Prior, D. J. 1985. Water regulatory behaviour in terrestrial gastropods. *Biol. Res.* 60:403–424.

Racki, G. 1982. Ecology of the primitive charophyte algae: A critical review. *Neues. Jahrb. Geol. Paläcont. Abh.* 162:388–399.

Radinsky, L. B. 1987. *The Evolution of Vertebrate Design*. Chicago: University of Chicago Press.

Ragan, M. A. and D. J. Chapman. 1978. *A Biochemical Phylogeny of Protists*. New York: Academic Press.

Rahn, H. and P. Dejours. 1987. Respiratory transition and acid-base balance from water to air. In P. Dejours, L. Bolis, C. R. Taylor, and E. R. Weibel, eds., *Comparative Physiology*, pp. 27–36. Padova, Italy: Liviana Press.

Ramshaw, J. A. M., D. Peacock, B. T. Meatyard et al. 1974. Phylogenetic implications of the amino acid sequence of cytochrome c from Enteromorpha intestinalis. *Phytochem.* 13:2783–2789.

Ramsköld, L. and H. Xianguang. 1991. New early Cambrian animal and onychophoran affinities of enigmatic metazoans. *Nature* 351:225–228.

Raubeson, L. A. and R. K. Jansen. 1992. Chloroplast DNA evidence on the ancient evolutionary split in vascular land plants. *Science* 255:1697–1699.

Raven, J. A. 1977. The evolution of vascular land plants in relation to supracellular transport processes. *Adv. Bot. Res.* 5:153–219.

Raven, J. A. 1984. Physiological correlates of the morphology of early vascular plants. *Bot. J. Linn. Soc.* 88:105–126.

—— 1985. Comparative physiology of plant and arthropod land adaptation. *Phil. Trans. Roy. Soc. London* B309:273–288.

Rayner, R. J. 1989. The puzzle of the first vascular plants: The South African connection. *South Afr. J. Sci.* 85:552–557.

Rees, B. B. and S. C. Hand. 1990. Heat dissipation, gas exchange and acid-base status in the land snail Oreohelix during short-term estivation. *J. Exp. Biol.* 152:77–92.

Regnault, M. 1987. Nitrogen excretion in marine and fresh-water Crustacea. *Biol. Rev.* 62:1–24.

Reilly, S. M. and G. V. Lauder. 1990. The evolution of tetrapod feeding behavior: Kinematic homologies in prey transport. *Evolution* 44:1542–1557.

Remy, W. and R. Remy. 1980a. Lyonophyton rhyniensis nov. gen. et nov. spec., ein Gametophyt aus dem Chert von Rhynie (Unterdevon, Schottland). *Argumenta Palaeobot.* 6:37–72.

—— 1980b. Devonian gametophytes with anatomically preserved gametangia. *Science* 208:295–296.

Remy, W., R. Remy., H. Hass et al. 1980. Sciadophyton Steinmann: Ein Gametophyt aus dem Siegen. *Argumenta Palaeobot.* 6:73–94.

Repetski, J. E. 1978. A fish from the Upper Cambrian of North America. *Science* 200:529–531.

Retallack, G. J. 1981. Fossil soils: Indicators of ancient terrestrial environments. In K. J. Niklas, ed., *Paleobotany, Paleoecology and Evolution*, pp. 55–102. New York: Praeger.

—— 1985. Fossil soils as grounds for interpreting the advent of large plants and animals on land. *Phil. Trans. Royal Soc. London* B309:105–142.

—— 1986. The fossil record of soil. In V. P. Wright, ed., *Paleosols, their Recognition and Interpretation*, pp. 1–57. Princeton, New Jersey: Princeton University Press.

Retallack, G. J. and C. R. Feakes. 1987. Trace fossil evidence for late Ordovician animals on land. *Science* 235:61–63.

Riddle, W. A. 1983. Physiological ecology of land snails and slugs. In W. D. Russell-Hunter, ed., *The Mollusca*, vol. 6: Ecology, pp. 431–461. Orlando, Fl.: Academic Press.

Rittschof, D. 1992. Chemosensation in the daily life of crabs. *Amer. Zoologist* 32:363–369.

Rittschof, D. and J. P. Sutherland. 1986. Field studies on chemically mediated behavior in land hermit crabs: Volatile and nonvolatile odors. *J. Chem. Ecology* 12:1273–1284.

Robertson, J. R., J. A. Fudge, and G. K. Vermeer. 1981. Chemical and live feeding stimulants of the sand fiddler crab, Uca pugilator (Bosc). *J. Exp. Mar. Biol. Ecol.* 53:47–64.

Robinson, P. L. 1973. Palaeoclimatology and continental drift. In D. H. Tarling and S. K. Runcorn, eds., *Implications to the Earth Sciences*, vol. 1, pp. 451–476. New York: Academic Press.

Robison, R. A. 1985. Affinities of Aysheaia (Onychophora), with description of a new species. *J. Paleontol.* 59:226–235.

——— 1990. Earliest-known uniramous arthropod. *Nature* 343:163–164.

Rohdendorf, B. B. and A. P. Rasnitsyn. 1980. Historical development of the Class Insecta. *Trans. Paleontol. Inst.* 175:1–270.

Rolfe, W. D. I. 1969. Arthropleurida and Arthropoda incertae sedis. In R. C. Moore, ed., *Treatise on Invertebrate Paleontology*, Part R, pp. 607–625. Geol. Soc. America.

——— 1980. Early invertebrate terrestrial faunas. In A. L. Panchen, ed., *The Terrestrial Environment and the Origin of Land Vertebrates*, Systematics Assoc. Spec. vol. 15, pp. 117–157. London: Academic Press.

——— 1985. Early terrestrial arthropods, a fragmentary record. *Phil. Trans. Royal Soc. London* B309:207–218.

——— 1990. Seeking the arthropods of Eden. *Nature* 348:112–113.

Rolfe, W. D. I., F. R. Schram, C. V. Pacaud et al. 1982. A remarkable Stephanian biota from Montceau-les-Mines, France. *J. Paleontol.* 56:426–428.

Romankiw, L. A., P. G. Hatcher, and G. B. Roen. 1988. Evidence of land plant affinity for the Devonian fossil Protosalvinia (Foerstia). *Lethaia* 21:417–423.

Rome, L. C., E. D. Stevens, and H. B. John-Alder. 1992. The influence of temperature and thermal acclimation on physiological function. In M. E. Feder and W. W. Burggren, eds., *Environmental Physiology of the Amphibians*, pp. 183–205. Chicago: University of Chicago Press.

Romer, A. S. 1960. Atemnospondylous labyrinthodont from the Lower Carboniferous. *Kirtlandia* 6:1–20.

Romer, A. S. and B. H. Grove. 1934. Environment of early vertebrates. *Amer. Midland Nat.* 16:805–856.

Rosen, D. E., P. L. Forey, B. G. Gardiner, and C. Patterson. 1981. Lungfishes, tetrapods, paleontology and plesiomorphy. *Bull. Am. Mus. Nat. Hist.* 167:159–276.

Rosenberg, G. D. and S. K. Runcorn. 1975. *Growth and Rhythms and the History of the Earth's Rotation*. New York: Wiley Interscience.

Roth, G. 1987. *Visual Behavior in Salamanders*. New York: Springer-Verlag.

Rouffa, A. S. 1978. On phenotypic expression, morphogenetic pattern, and synangium evolution in Psilotum. *Amer. J. Bot.* 65:692–713.

Sacca, R. and W. Burggren. 1982. Oxygen uptake in air and water in the airbreathing reedfish Calamoichthys calabaricus: Role of skin, gills and lungs. *J. Exp. Biol.* 97:179–186.

Salvini-Plawen, L. V. 1987. The structure and function of molluscan digestive systems. In E. R. Trueman and M. R. Clarke, eds., *The Mollusca*,

294   References

vol. 11: Form and Function, pp. 301–379. Orlando, Fl.: Academic
Press.
Sansom, I. J., M. P. Smith, H. A. Armstrong et al. 1992. Presence of the
earliest vertebrate hard tissues in conodonts. *Science* 256:1308–1311.
Sawyer, R. T. 1986. *Biology of Leeches. Vol. I: Anatomy, Physiology, and Behav-
iour. Vol. II: Feeding Biology, Ecology and Systematics. Vol. III: Bibliography.*
Oxford: Oxford University Press.
Sayer, M. D. J. and J. Davenport. 1991. Amphibious fish: Why do they leave
water? *Rev. Fish Biol. Fisheries* 1:159–181.
Scagel, R. E., R. J. Bandoni, J. R. Maze et al. 1984. *Non-vascular Plants.*
Belmont, Calif.: Wadsworth.
Schmid, R. 1976. Septal pores in Prototaxites, an enigmatic Devonian plant.
Science 191:287–288.
Schram, F. R. 1978. Arthropods: A convergent phenomenon. *Fieldiana: Geol-
ogy* 39:61–108.
—— 1986. *Crustacea.* New York: Oxford University Press.
Schultze, H.-P. 1981. Hennig und der Ursprung der Tetrapoda. *Paleontolog.
Zeits.* 55:71–86.
Schultze, H.-P. and L. Trueb, eds. 1991. *Origins of the Higher Groups of
Tetrapods: Controversy and Consensus.* Ithaca: Cornell University Press.
Schuster, R. M. 1983. Evolution, phylogeny and classification of the Hepati-
cae. In R. M. Schuster, ed., *A New Manual of Bryology,* vol. 2., pp. 893–
1070. Nichinan, Japan: Hattori Botanical Lab.
Scotese, C. R., R. K. Bambach, C. Barton et al. 1979. Paleozoic base maps. *J.
Geol.* 87:217–277.
Scotese, C. R. and W. S. McKerrow. 1990. Revised world maps and introduc-
tion. In W. S. McKerrow and C. R. Scotese, eds., *Palaeozoic Paleogeography
and Biogeography, Memoir 12,* pp. 1–10. Geol. Soc. London.
Seilacher, A. 1989. Vendozoa: Organismic construction in the Proterozoic
biosphere. *Lethaea* 22:229–239.
Selden, P. 1981. Functional morphology and the prosoma of Baltoeurypterus
tetragonophthalmus Fisher (Chelicerata: Eurypterida). *Trans. Roy. Soc.
Edinburgh* 72:9–48.
—— 1984. Autecology of Silurian eurypterids. *Spec. Papers Paleontol.* 32:39–
54.
Selden, P. A. and D. Edwards. 1990. Colonisation of the land. In K. C.
Allen and D. E. Briggs, eds., *Evolution and the Fossil Record,* pp. 122–152.
Washington, D.C.: Smithsonian Institution Press.
Semlitsch, R. D. and H. M. Wilbur. 1988. Effects of pond drying time on
metamorphosis and survival in the salamander Ambystoma talpoideum.
Copeia 4:978–983.
Seymour, R. S. and J. D. Roberts. 1991. Embryonic respiration and oxygen
distribution in foamy and nonfoamy egg masses of the frog Limnody-
nastes tasmaniensis. *Physiol. Zool.* 64:1322–1340.
Shaffer, H. B. and F. Breden. 1989. The relationship between allozyme

variation and life history: Non-transforming salamanders are less variable. *Copeia* 4:1016–1023.

Shaffer, H. B. and G. V. Lauder. 1988. The ontogeny of functional design: Metamorphosis of feeding behaviour in the tiger salamander (Ambystoma tigrinum). *J. Zool. London* 216:437–454.

Sharov, A. G. 1957. Original Paleozoic wingless insects of the new Order Monura (Insecta: Apterygota). *Dokl. Acad. Sci. U.S.S.R* 115:795–798.

Shear, W. A. 1990. Silurian-Devonian terrestrial arthropods. *Short Courses Paleontology* 3:197–213.

—— 1991. The early development of terrestrial ecosystems. *Nature* 351:283–289.

Shear, W. A., P. M. Bonamo, J. D. Grierson et al. 1984. Early land animals in North America: Evidence from Devonian age arthropods from Gilboa, New York. *Science* 224:492–494.

Shear, W. A. and J. Kukalova-Peck. 1990. The ecology of Paleozoic terrestrial arthropods: The fossil evidence. *Can. J. Zool.* 68:1807–1834.

Shear, W. A., J. M. Palmer, J. A. Coddington et al. 1989. A Devonian spinneret: Early evidence in the fossil record for spiders and their use of silk. *Science* 246:479–481.

Sherwood-Pike, M. A. and J. Gray. 1985. Silurian fungal remains: Probable records of the class Ascomycetes. *Lethaia* 18:1–20.

Shoemaker, V. H., M. A. Baker, and J. P. Loveridge. 1989. Effect of water balance on thermoregulation in waterproof frogs (Chiromantis and Phyllomedusa). *Physiol. Zool.* 62:133–146.

Shoemaker, V. H., S. S. Hillman, S. D. Hillyard et al. 1992. Exchange of water, ions and respiratory gases in terrestrial amphibians. In M. E. Feder and W. W. Burggren, eds., *Environmental Physiology of the Amphibians*, pp. 125–150. Chicago: University of Chicago Press.

Simkiss, K. 1987. Molluscan skin (excluding cephalopods). In E. R. Trueman and M. R. Clarke, eds., *The Mollusca*, vol. 11: Form and Function, pp. 11–35. Orlando, Fl.: Academic Press.

Skinner, D. M. 1985. Molting and regeneration. In D. E. Bliss and L. H. Mantel, eds., *The Biology of Crustacea*, vol. 9: Integument, Pigments and Hormonal Processes, pp. 43–146. Orlando, Fl.: Academic Press.

Sluiman, H. J. 1988. A cladistic evaluation of the lower and higher green plants. *Pl. System. Evol.* 149:217–232.

Smith, A. J. E. 1986. Bryophyte phylogeny: Fact or fiction. *J. Bryol.* 14:83–89.

Smithson, T. R. 1989. The earliest known reptiles. *Nature* 342:676–678.

Snodgrass, R. E. 1938. The evolution of Annelida, Onychophora and Arthropoda. *Smithsonian Misc. Coll.* 97:1–59.

Solem, A. 1979. Biogeographic significance of land snails, Paleozoic to Recent. In J. Gray and A. Boucot, eds., *Historical Biogeography, Plate Tectonics and the Changing Environment*, pp. 277–287. Corvallis: Oregon State University Press.

—— 1985. Origin and diversification of pulmonate land snails. In E. R.

Trueman and M. R. Clarke, eds., *The Mollusca*, vol. 10: Evolution, pp. 269–293. Orlando, Fl.: Academic Press.

Solem, A. and E. L. Yochelson. 1979. *North American Paleozoic land snails with a summary of other Paleozoic non-marine snails*. Prof. Paper U. S. Geol. Surv. 1072.

Somero, G. M. 1987. Organic osmolyte systems: Convergent evolution in the design of the intracellular milieu. In P. D. L. Bolis, C. R. Taylor, and E. R. Weibel, eds., *Comparative Physiology*, pp. 207–218. Padova: Liviana Press.

Speck, T. and D. Vogellehner. 1988. Biophysical examinations of the bending stability of various stele types and the upright axes of early "vascular" land plants. *Bot. Acta* 101:262–267.

Spicer, J. I., P. G. Moore, and A. C. Taylor. 1987. The physiological ecology of land invasion by the Talitridae (Crustacea: Amphipoda). *Proc. Roy. Soc. London* B232:95–124.

Spicer, J. I. and A. C. Taylor. 1987. Respiration in air and water of some semi- and fully terrestrial talitrids (Crustacea: Amphipoda: Talitridae). *J. Exp. Mar. Biol. Ecol.* 106:265–277.

Spinar, Z. V. 1952. Revise nekterych moravsky'ch Diskosauridu (Revision of some Moravian Discosauriscidae). *Roz. Ustred. Ustav. Geol.* 14:1–159 (Czech with English summary).

Spjeldnaes, N. 1979. The palaeoecology of the Ordovician Harding Sandstone. *Palaeogeog. Palaeoclimatol. Palaeoecol.* 26:317–347.

Stebbins, G. L. and G. J. C. Hill. 1980. Did multicellular plants invade the land. *Amer. Naturalist* 115:342–353.

Steinbrecht, R. A. 1984. Chemo-, hygro-, and thermoreceptors. In J. Bereiter-Hahn, A. G. Matoltsy, and K. S. Richards, eds., *Biology of the Integument*, vol. 1: Invertebrates, pp. 523–553. Berlin: Springer-Verlag.

Stevenson, J. R. 1985. Dynamics of the integument. In D. E. Bliss and L. H. Mantel, eds., *The Biology of Crustacea*, vol. 9: Integument, Pigments, and Hormonal Processes, pp. 1–42. Orlando, Fl.: Academic Press.

Stewart, K. D. and K. R. Mattox. 1975. Comparative cytology, evolution and classification of the green algae, with some consideration of the origin of other organisms with chlorophylls a and b. *Bot. Rev.* 41:104–135.

—— 1978. Structural evolution in the flagellated cells of green algae and land plants. *Biosystems* 10:145–152.

Stewart, W. N. 1983. *Paleobotany and the Evolution of Plants*. Cambridge: Cambridge University Press.

Stiffler, D. F., M. L. D. Ruyter, and C. R. Talbot. 1990. Osmotic and ionic regulation in the aquatic caecilian Typhlonectes compressicauda and the terrestrial caecilian Ichthyophis kohtaoensis. *Physiol. Zool.* 63:649–668.

Storey, K. B. and J. M. Storey. 1988. Freeze tolerance in animals. *Physiol. Rev.* 68:27–84.

—— 1992. Natural freeze tolerance in ectothermic vertebrates. *Ann. Rev. Physiol.* 54:619–637.

Størmer, L. 1976. Arthropods from the Lower Devonian (Lower Emsian) of Alken an der Mosel, Germany, pt. 5. Myriapods and additional forms, with general remarks on fauna and problems regarding invasion of land by arthropods. *Senkenberg Lethaea* 57:87–183.

—— 1977. Arthropod invasion of the land during late Silurian and Devonian times. *Science* 197:1362–1364.

Strother, P. K. 1988. A new species of Nematothallus from the Silurian Bloomsburg formation of Pennsylvania. *J. Paleont.* 62:967–982.

Stubblefield, S. P. and T. N. Taylor. 1988. Recent advances in paleomycology. *New Phytol.* 108:3–25.

Swain, T. and G. Cooper-Driver. 1984. Biochemical evolution in early land plants. In K. J. Niklas, ed., *Paleobotany, Paleoecology and Evolution*, pp. 103–134. New York: Praeger.

Takeda, N. 1984. The aggregation phenomenon in terrestrial isopods. *Symp. Zool. Soc. London* 53:381–404.

Taylor, E. W., N. M. Whiteley, and M. G. Wheatly. 1991. Respiratory gas exchange and the regulation of acid-base status in decapodan crustaceans. In A. J. Woakes, M. K. Grieshaber, and C. R. Bridges, eds., *Physiological Strategies for Gas Exchange and Metabolism*, pp. 79–106. Cambridge: Cambridge University Press.

Taylor, T. N. 1988. The origin of land plants: Some answers, more questions. *Taxon* 37:805–833.

Theriot, E. 1988. A review of Sluiman's cladistic classification of green plants with particular reference to flagellar data and to land plant origins. *Taxon* 37:913–919.

Thomas, B. A. and R. A. Spicer. 1986. *The Evolution and Palaeobiology of Land Plants*. Portland, Ore.: Dioscorides Press.

Thompson, I. and D. S. Jones. 1980. A possible onychophore from the Middle Pennsylvanian Mazon Creek Beds of Northern Illinois. *J. Paleontol.* 54:588–596.

Tiegs, O. W. and S. M. Manton. 1958. The evolution of Arthropoda. *Biol. Rev.* 32:255–337.

Toews, D. P. and D. MacIntyre. 1977. Blood respiratory properties of a viviparous amphibian. *Nature* 266:464–465.

—— 1978. Respiration and circulation in an apodan amphibian. *Can. J. Zool.* 56:998–1004.

Tompa, A. S. 1984. Land snails (Stylommatophora). In A. S. Tompa, N. H. Verdonck, and J. A. M. van den Biggelaar, eds., *The Mollusca*, vol. 7: Reproduction, pp. 47–140. Orlando, Fl.: Academic Press.

Townsend, D. S. and M. M. Stewart. 1985. Direct development in Eleutherodactylus coqui (Anura: Leptodactylidae): A staging table. *Copeia* 2:423–436.

Trott, T. J. and J. R. Robertson. 1984. Chemical stimulants of cheliped flexion behavior by the western Atlantic ghost crab Ocypode guadrata (Fabricius). *J. Exp. Mar. Biol. Ecol* 78:237–252.

Truchot, J. P. 1987. *Comparative Aspects of Extracellular Acid-Base Balance*. Berlin: Springer-Verlag.

—— 1990. Respiratory and ionic regulation in invertebrates exposed to both water and air. *Ann. Rev. Physiol.* 52:61–76.

Trueman, E. R. 1983. Locomotion in molluscs. In A. S. M. Saleuddin and K. M. Wilbur, eds., *The Mollusca*, vol. 4: Physiology, Part 1, pp. 155–198. New York: Academic Press.

Trueman, E. R. and M. R. Clarke. 1988. Introduction. In E. R. Trueman and M. R. Clarke, eds., *The Mollusca*, vol. 11: Form and Function, pp. 1–9. San Diego: Academic Press.

Uchiyama, M., T. Murakami, and H. Yoshizawa. 1990. Notes on the development of the crab-eating frog, Rana cancrivora. *Zool. Sci.* 7:73–78.

Valentine, J. W. 1989. Bilaterians of the Precambrian-Cambrian transition and the annelid-arthropod relationship. *Proc. Nat. Acad. Sci. USA* 86:2272–2275.

Vandel, A. 1943. Essai sur l'origine et l'evolution et la classification des Oniscoides (Isopodes terrestres). *Bull. Biol. France Belge (Supplement)* 30:1–143.

—— 1965. Sur l'existence d'oniscoides tres primitifs menant une vie aquatique et sur le polyphyletisme des isopodes terrestres. *Ann. Speleologie* 20:485–518.

Vitalis, T. Z. and G. Shelton. 1990. Breathing in Rana pipiens: The mechanism of ventilation. *J. Exp. Biol.* 154:537–556.

Wake, D. B. and G. Roth, eds. 1989. *Complex Organismal Functions*. New York: Wiley Interscience.

Walcott, C. D. 1892. Notes on the discovery of a vertebrate fauna in Silurian (Ordovician) strata. *Bull. Geol. Soc. Amer.* 3:153–172.

—— 1911. Middle Cambrian Merostomata; Middle Cambrian Holothuria and Medusae; Middle Cambrian annelids. In *Cambrian Geology and Paleontology II*, Smithsonian Misc. Coll., vol. 57, pp. 17–40, 41–68, 109–144.

Walsh, P. J. and R. P. Henry. 1991. Carbon dioxide and ammonia metabolism and exchange. In P. W. Hochachka and T. P. Mommsen, eds., *Biochemistry and Molecular Biology of Fishes*, vol. 1, pp. 181–207. Amsterdam: Elsevier.

Walter, M. 1979. Precambrian glaciation. *Amer. Scientist* 67:142.

Walton, M. and B. D. Anderson. 1988. The aerobic cost of saltatory locomotion in the Fowler's toad (Bufo woodhousei fowleri). *J. Exp. Biol.* 136:273–288.

Warburg, M. R. 1987. Isopods and their terrestrial environment. *Adv. Ecol. Res.* 17:187–242.

Warren, A. B., R. Jupp, and B. Barrie. 1986. The earliest Devonian tetrapod trackway. *Alcheringa* 10:183–186.

Warren, J. W. and N. A. Wakefield. 1972. Trackways of tetrapod vertebrates from the Upper Devonian of Victoria, Australia. *Nature* 238:141–145.

Wassersug, R. J. 1989. Locomotion in amphibian larvae (or "Why aren't tadpoles built like fishes?"). *Amer. Zoologist* 29:65–84.

Waterson, C. D. 1975. Gill structures in the Lower Devonian eurypterid Tarsopterella scotia. *Fossil Strata* 4:241–245.

Wellins, C. A., D. Rittschof, and M. Wachowiak. 1989. Location of volatile odor sources by ghost crab Ocypode guadrata (Fabricius). *J. Chem. Ecology* 15:1161–1169.

Wells, J. W. 1963. Coral growth and geochronometry. *Nature* 197:948–950.

—— 1970. Problems of the annual and daily growth rings in corals. In S. K. Runcorn, ed., *Paleogeophysics*, pp. 3–9. London: Academic Press.

Wellstead, C. F. 1982. A Lower Carboniferous aistopod amphibian from Scotland. *Palaeontology* 25:193–208.

Westoll, T. S. I. e. 1977. Northern Britain. In R. M. House, ed., *A Correlation of the Devonian rocks in the British Isles*, vol. 8, pp. 66–93. Spec. Rep. Geol. Soc. London.

Wever, E. G. 1985. *The Amphibian Ear*. Princeton, New Jersey: Princeton University Press.

Whalley, P. and E. A. Jarzembowski. 1981. A new assessment of Rhyniella, the earliest known insect from the Devonian of Rhynie, Scotland. *Nature* 291:317.

Wheatly, M. G. and R. P. Henry. 1992. Extracellular and intracellular acid-base regulation in crustaceans. *J. Exp. Zool.* 263:127–142.

Whittington, H. B. 1978. The lobopod animal Aysheaia pedunculata Walcott, Middle Cambrian Burgess Shale, British Columbia. *Phil. Trans. Roy. Soc. London* B284:165–197.

—— 1985. *The Burgess Shale*. New Haven: Yale University Press.

Wieser, W. 1984. Ecophysiological adaptations of terrestrial isopods: A brief review. *Symp. Zool. Soc. London* 53:247–265.

Wilczynski, W. 1992. The nervous system. In M. E. Feder and W. W. Burggren, eds., *Environmental Physiology of the Amphibians*, pp. 9–39. Chicago: University of Chicago Press.

Wilkens, J. L. and R. E. Young. 1992. Regulation of pulmonary blood flow and of blood pressure in a mangrove crab (Goniopsis cruentata). *J. Exp. Biol.* 163:297–316.

Willows, A. O. D., ed. 1985. *The Mollusca*, vol. 8: Neurobiology and Behavior, Part 1. Orlando, Fl.: Academic Press.

Wills, L. J. 1965. Über die organization des Eurypterus fischeri Eichwald. *Arkiv für Zoologie* 18:93–145.

Wilson, A. C., H. Ochman, and E. M. Prager. 1987. Molecular time scale of evolution. *TIG* 3:241–247.

Wolcott, D. L. 1991. Nitrogen excretion is enhanced during urine recycling in two species of terrestrial crab. *J. Exp. Zool.* 259:181–187.

Wolcott, T. G. 1988. Ecology. In W. W. Burggren and B. R. McMahon, eds.,

Biology of the Land Crabs, pp. 55–96. Cambridge: Cambridge University Press.

—— 1992. Water and solute balance in the transition to land. *Amer. Zoologist* 32:428–437.

Wolcott, T. G. and D. L. Wolcott. 1991. Ion conservation by reprocessing of urine in the land crab Gecarcinus lateralis (Freminville). *Physiol. Zool.* 64:344–361.

Wollmuth, L. P. and L. J. Crawshaw. 1988. The effect of development and season on temperature selection in bullfrog tadpoles. *Physiol. Zool.* 61:461–469.

Wood, C. M., R. S. Munger, and D. P. Toews. 1989. Ammonia, urea and H+ distribution and the evolution of ureotelism in amphibians. *J. Exp. Biol.* 144:215–233.

Wood, S. C. and M. L. Glass. 1991. Respiration and thermoregulation of amphibians and reptiles. In A. J. Woakes, M. K. Grieshaber, and C. R. Bridges, eds., *Physiological Strategies for Gas Exchange and Metabolism,* pp. 107–124. Cambridge: Cambridge University Press.

Wood, S. P., A. L. Panchen, and T. R. Smithson. 1985. A terrestrial fauna from the Scottish Lower Carboniferous. *Nature* 314:355–356.

Wootton, R. J. 1981. Paleozoic Insects. *Ann. Rev. Entomol.* 26:319–344.

Wright, J. C. and J. Machin. 1990. Water vapour absorption in terrestrial isopods. *J. Exp. Biol.* 154:13–30.

Yoshiyama, R. M. and J. J. Cech. 1994. Aerial respiration by rocky intertidal fishes of California-Oregon. *Copeia* 1994 (1): 153–158.

Young, G. C. 1990. Devonian vertebrate distribution patterns and cladistic analysis of paleogeographic hypotheses. In W. S. McKerrow and C. R. Scotese, eds., *Palaeozoic Palaeogeography and Biogeography,* pp. 243–255. Geol. Soc. Mem. No. 12.

Zechman, F. W., E. C. Theriot, E. A. Zimmer et al. 1990. Phylogeny of the Ulvophyceae (Chlorophyta): Cladistic analysis of nuclear encoded rRNA sequence data. *J. Phycol.* 26:700–710.

Zeigler, A. M., C. R. Scotese, W. S. McKerrow et al. 1979. Paleozoic Paleogeography. *Ann. Rev. Earth Planetary Sci.* 7:473–502.

Zimmerman, W. 1953. Main results of the "Telome theory." *Palaeobotanist* 1:456–470.

General Index

Systematic Index of Animals

Latzelia, 122, 143
Lepospondyli, 100, 150
Lethiscus, 150
Ligia, 202–6, 209, 212
Ligidium, 205
Ligiidae, 212
Limulus, 125
Lissamphibia, 100, 149, 219, 254
Loxoma, 128

Malacostraca, 99, 139, 177
Mammalia, 100, 153
Mazonoscolopendra, 143
Melampus, 163
Merostomes, 108–10
Mesogastropoda, 98
Metaxygnathus, 129
Mictyris, 196
Mollusca, 96, 98, 102, 145–47, 160–75
Moniligastrida, 98
Monura, 144
Myriapoda, 99, 111, 113, 119, 120, 142, 143–44, 254
Mystacocarida, 138

Natantia, 141
Necrogammarus, 113
Nematoda, 96, 98
Nemertinea, 96, 98
Neoptera, 144

Ocypode, 210
Oligochaeta, 98, 108
Onchidium, 163
Oniscoidea, 140
Onychophora, 99, 103, 105, 142
Opiliones, 134
Orconectes, 193–94
Osmeridae, 240
Osteichthyes, 99, 218–19
Osteolepidae, 131
Ostracoda, 99, 106, 113, 124, 139
Otala, 163

Pachygrapsus, 188
Palaeocharinoides, 122
Palaeocharinus, 122, 123
Palaeoctenzia, 123
Palaeophonus, 122
Palaeoptera, 144
Palpigradi, 137
Parahughmilleria, 124
Pauropoda, 143
Peracarida, 139, 177
Perciformes, 219, 240, 245–47
Periophthalmus, 245
Peripatus, 103, 143
Phyllodocemorpha, 98
Pikaia, 103, 107
Platyhelminthes, 96, 98
Polychaeta, 98, 104, 106
Polyplacophora, 160
Polypterus, 226
Pomatiasidae, 146
Procambarus, 187
Prosobranchia, 98, 146, 161, 162–63, 165–66, 170–71, 172–74, 253, 255
Protacaris, 122, 123
Pulmonata, 98, 146–47, 161, 162–65, 166–69, 170–72, 172–74, 253, 255
Pupidae, 146

Rana, 236
Reptantia, 141
Reptilia, 100, 152–56
Rhipidistia, 99, 127, 129, 254, 264
Rhynchobdellida, 98, 177
Rhyniella, 122, 123
Romeriscus, 153

Sarcopterygia, 218
Sauropsida, 100, 153
Schizomida, 137
Scopimera, 196
Scorpiones, 98, 113, 119, 124, 253, 255
Serracaris, 105
Seymouria, 152

Systematic Index of Plants